Unlocking the Atom

The Canadian Book
on Nuclear Technology

Hans Tammemagi

David Jackson

mcmaster university press

Published in 2002 by McMaster University Press
For further information or to order additional copies contact:
www.unlockingtheatom.ca

National Library of Canada Cataloguing in Publication Data

Tammemagi, H. Y.
Unlocking the atom: the Canadian book on nuclear technology

Includes bibliographical references and index.
ISBN 0-9730040-0-2

1.   Nuclear energy—Popular works.
2.   Nuclear engineering—Popular works.
3.   Nuclear energy—Canada—Popular works.
4.   Nuclear industry—Canada—Popular work. I. Jackson, David (David Phillip), 1940- II. Title.

TK9146.T34 2001   333.792'4   C2001-904077-6

CANDU® and CANFLEX® are registered trademarks of Atomic Energy of Canada Limited.
Ballard® is a registered trademark of Ballard Power Systems Ltd.

The figures from the Canadian Nuclear Safety Commission Publication "Canada: Living with Radiation" are reproduced with the permission of the Minister of Public Works and Government Services Canada.

Every reasonable effort has been made to contact the holders of copyright for materials reprinted in this book. The authors will gladly receive information that will enable them to rectify any inadvertent errors or omissions in subsequent editions. Please contact them at:

www.unlockingtheatom.ca

Cover photo: The core of the McMaster Nuclear Reactor. The blue glow is Cerenkov radiation from relativistic electrons passing through water. *Photo credit:* McMaster University

Printed and Bound in Canada

Contents

Responsibility
Regulatory Philosophy
CNSC Structure
Licensing Process
Nuclear Substances and Radioactive Materials
Transportation
Nuclear Liability Insurance
Safeguards
Other Canadian Jurisdictions
Exposure Limits
Nuclear Critics

Acid Rain and Air Pollution
Global Warming
Reactor Emissions
Overview of Nuclear Wastes
Toxicity
Quantities
Duration of Toxicity
Disposal Methods
Low-Level Radioactive Waste
The Bruce Nuclear Power Development Site
Chalk River
Historical Wastes
Port Hope, Ontario
Other Historical Wastes in Canada
The Low-Level Radioactive Waste Management Office
The Swedish Approach to LLRW Disposal

Nuclear Fuel Wastes
Interim Storage
Permanent Disposal
Regulation
Canadian Environmental Assessment Agency Panel
Review
Summary

Diagnostic Nuclear Medicine
X-rays
Radioactive Tracers

Foreword

I remember standing at Land's End, the westernmost point of England, amongst granite crags that towered over me like timeless foreboding hulks. Waves crashed against the rocky cliffs, as they have done for eternity, rolling in from far out in the wild reaches of the Atlantic to cast themselves upon this desolate coast. Behind me the barren moor rolled in gentle undulations to the far horizon. The occasional road, bordered by dark stone hedges, wandered through the bleakness as though searching for a hospitable place.

What lured me to this stark yet beautiful place were the rocks. My students and I spent several weeks scrambling onto high crags with strange names such as Rough Tor, Trewavas Head, Carn Brea, and Brown Willy. We climbed into quarries and dropped down the black shafts of tin mines far into the earth, with the sole objective of retrieving samples of granite. After long days on the moors, armed with pints of ale, we hunched over our day's data in pubs with ancient beams and smoke-stained walls erected long before Canada became a nation. The bleak Cornish landscape formed a wonderfully romantic setting for my introduction to the topic of radioactivity.

As a research fellow at Imperial College, London, I was investigating the radioactivity of southwest England. Sensitive instruments at the University of London's research reactor site allowed me to measure the concentrations of potassium, uranium, and thorium in the rock samples. Mapping the distribution of these radioactive elements provided clues on the heat flowing from the earth and how the geology of the area formed. As we discovered, the moors of Cornwall and Devon are one of the most radioactive areas of Great Britain.

As part of my geophysical studies, I learned that radiation has been a fundamental part of the environment since the earth was formed. In fact, energy from the radioactivity of the earth is the driving force for tectonic plates, large continent-sized slabs that move about on the earth's surface. Where they collide, the plates form majestic mountain ranges like the Himalayas, the Andes, and the Rockies. As erosion wears away old mountains, new ones are being continually created by this dynamic process. Thus, natural radioactivity is directly responsible for the topography and beauty of the globe — without it the earth would have eroded long ago into a boring, featureless sphere totally covered by oceans.

The radioactivity from rocks enters soils and building materials becoming an inherent part of the homes we live in. Radioactivity is taken up by plants through their roots and then enters animals and humans as food.

Rocks are not the only source of natural radioactivity. The sun, like all stars, is a giant nuclear reactor, constantly sending out streams of radiation. This solar "wind" strikes the earth's atmosphere where some of the radiation is absorbed and some passes through to ground level.

Thus, all rocks, all soils, all plants, and all living creatures are immersed in a sea of natural radioactivity, and have been since the beginning of time.

Hans Tammemagi

Preface

Canada has a long and rich history in the nuclear field. Nuclear technologies of all kinds are embedded in the history and fabric of our society. Consider the following:

• Canada was just the second country to build an operating nuclear reactor.

• Canada has developed its own unique electricity-producing nuclear reactor, the CANDU®, which is competitive in the world market.

• Canadians are among the world's largest consumers of nuclear electricity and have amongst the highest number of nuclear plants per capita of any nation in the world.

• Canada is the world's largest uranium exporter and has the world's largest proven reserves of uranium.

• Canada is the world's largest producer and exporter of radioactive materials for medical diagnosis and treatment. About 12 million medical procedures are conducted each year around the world using medical isotopes made in Canada.

• Canada is a world leader in the manufacture of radiation devices for medicine and industry including cancer therapy machines, food irradiators, sterilizers for medical products, and irradiators for industrial processes.

In spite of nuclear technology and radioactivity being such a pervasive part of our lives and our natural surroundings, they remain highly controversial subjects. Nuclear power is one of those technologies, like cloning and genetically modified foods, that attracts public opposition and passion. Even small nuclear incidents command headlines. Some view nuclear power as a dangerous source of electricity. Opposition is routinely mounted against all proposed nuclear projects whether they are power plants, waste disposal facilities, or food irradiation centres.

On the other side, the nuclear industry, and sometimes governments, view opposition to things nuclear as an emotional reaction, rooted in irrational fears. Since the nuclear scientists and engineers are the experts, they adopt a "we know best" attitude in the face of opposition.

For some, nuclear technology is a symbol for many of the problems in modern society. A concern for the environment, assuring a sustainable and healthy future, the maintenance of world peace, the control of science and technology by citizens, distrust of big companies and governments, and a longing for a self-sufficient and simple lifestyle all enter into the controversy. In making decisions on complex technologies, both social as well as scientific issues must be considered.

A common feature in the nuclear debate is the hurling of charges and countercharges by opponents and proponents. Many of these missiles are unguided in terms of the information they contain and the interpretation of that information. This lack of valid content in much of what is being said is counter-productive. For the nuclear debate to have any value, people need to communicate using facts. To date, this has been lacking in Canada and this book is intended to remedy that situation.

Canada has been a nuclear nation for over fifty years and yet there has been no book that provides information that a layperson can understand. There have been a few academic histories of nuclear development and some technical books on specialized aspects of nuclear technology but, surprisingly, there is not even an undergraduate university textbook. The nuclear industry has published many pamphlets over the years, as have various anti-nuclear groups. Needless to say, such publications are not comprehensive and certainly not unbiased.

This book is unashamedly Canadian. Wherever we can, we highlight the many nuclear discoveries, inventions, and firsts that Canada has produced and the often extraordinary men and women who accomplished them. Whatever one might feel about nuclear technology itself, Canadian achievements in this field have been significant and should be recognized as part of our national heritage.

We believe there is a need for factual information on nuclear subjects and that Canadians are capable of making sound decisions about these matters. This book provides the information necessary for informed decisions in as unbiased a manner as we are able.

Using plain language, we explain the fundamentals of nuclear reactions and radioactivity and how the various nuclear technologies work. We have tried to make these explanations simply and without sacrificing accuracy. All facets of nuclear technology are explained from uranium mining to electricity-producing nuclear power reactors, including the Canadian-designed CANDU reactor, to nuclear medicine and industrial applications. The history of the development of nuclear technology in Canada is outlined. The health impacts of radiation and the technical controversy that surrounds this topic are also described.

Nuclear power plants are compared to other sources of energy and nuclear wastes are compared to other types of waste created in society. You will be able to assess nuclear technology in relation to its alternatives. Appendix A contains an explanation of radiation and nuclear processes. New words and acronyms are bolded the first time they appear in the text and are explained in the Glossary.

This book is intended for those in the general public who wish to learn about nuclear science and technology. It is presented in an easy-to-read format and is illustrated by many figures and photographs. A few chapters will require some high-school science and a bit of mathematics, but the main prerequisite is an interest in learning about the subject. Because it is relatively comprehensive, this book will also be a useful reference in Canadian universities, colleges and high-schools.

We hope this book will be a resource for anyone who wishes to gain an understanding of nuclear technology in the Canadian context.

Chapter One

Nuclear Technology at the Crossroads

There is something about the word 'nuclear' that causes the pulse to beat faster and the blood pressure to rise. The mere mention of the word is guaranteed to raise emotions and liven up the dullest party. Some people loathe anything connected with nuclear and oppose it fiercely. Others find nuclear technology acceptable. Many, perhaps the majority, don't know much about the topic other than what they hear through the media. They worry that a major reactor accident, such as at Chernobyl, could happen in Canada, and that radioactivity, even in small quantities, can cause damage to health; this is particularly frightening as radioactivity is invisible and is thought to cause genetic deformities. They also worry that there is no safe method of disposing of nuclear wastes, which are seen as one of the most toxic substances ever produced by humans.

However you view it, the debate that swirls around nuclear issues is vitally important. Similar uncertainties and fears, although perhaps not to the same degree, surround other technologies such as genetic modification of food, incineration of municipal waste, and the manufacture and application of pesticides. How does society control powerful technologies that have the potential for both harm and good? What mechanisms do we set in place to ensure we reap the benefits while at the same time minimizing or eliminating harm to human health and the environment? How do we make appropriate decisions, and enforce them?

1

Finding answers to these questions is essential, as technology is critical to the survival of the human race. Let us explore why.

Today, humans are teeming on this globe. In 1999, the earth's population reached six billion — an astonishing seven-fold increase in the past two centuries. And the beat goes on. For the next 15 years, global population growth is estimated to be over 80 million people per year — another billion souls every 12.5 years. The United Nations calculates that by the year 2050 the population could be 11 billion people — almost double what we have now! That is an enormous number of people, working, eating, jostling, purchasing, competing with one another, and all seeking a better way of life.

As early as 1800, Thomas Malthus, a demographer who studied population growth, predicted that human population was increasing so rapidly it would soon outstrip the resources of the earth. The horror of famine and catastrophe would inevitably follow. Many others have made the same prediction since that time.

But the Malthusian Devil never arrived. Remarkably, the production of food — not to mention a cornucopia of other consumer goods — has kept pace with population growth, and the standard of living for most people has improved significantly. We live better and longer than any of our ancestors. We have more choice, more ability to travel, more information, better medical support and education, more holidays, and more income.

So why was Malthus wrong? Very simply, he vastly underestimated the enormous power of human ingenuity. Science and technology have done wonders in providing increasing quantities of life's necessities. Plant genetics, fertilizers, and pesticides have increased crop yields enormously; synthetic materials and weaving machines have led to mass-produced clothes; and Henry Ford's production line has manufactured countless cars and other consumer goods. For two centuries the Malthusian Devil has been kept at bay, thanks to technology, our tool-making ability.

But there has been a price to pay. The sheer weight of human numbers, combined with the inappropriate use of many technologies, is slowly but surely overwhelming the natural ecosystems of the earth.

Everywhere we look there is pollution: in the air we breathe, the water we drink, in lakes and rivers, and in soils and groundwater.

Pollution is damaging our health (more than 5000 deaths a year in Canada due to air pollution alone), degrading our quality of life, and damaging the resources we need for industry.

Global warming, caused largely by burning massive quantities of fossil fuels, will cause dramatic weather changes that will flood low-lying coastal areas and could turn the prairie bread basket into a dust bowl. And there is so much more: crowded cities are incubators for powerful new diseases such as AIDS, *E. Coli* 1057:H7, and the Ebola virus to name just a few. Aquifers are being drained or contaminated, old-growth and tropical forests are being razed, top-soil is becoming polluted or is blowing away, and an enormous pool of biodiversity is being wiped out.

The saddest result of our unfettered proliferation is the extinction of other species. Claiming the top rung in the animal kingdom, humans should be the custodian species protecting creatures small and large; instead, *Homo sapiens* would more accurately be described as the exterminator species. Instead of benignly ruling the natural kingdom, we ride roughshod over it.

With the population increasing at its current rates, we have no choice but to use advanced technologies — and continue to develop new ones — if we wish to feed, clothe, shelter and provide a satisfactory standard of living. Returning to a simpler way of life simply is not an option. But we must learn to use these technologies more intelligently than in the past. We must stop polluting the globe and stripping its resources. We must leave something for our grandchildren.

Nuclear technology is only another tool, albeit a powerful and complex one. It has enormous potential to bring benefits. If properly harnessed, it can provide an almost limitless source of clean energy, and it is making significant contributions to medical practice and the improvement of human health. It also has a dark side, however, with enormous destructive potential. But every tool that mankind has ever invented can be misused. For example, the manufacture of poisonous chemicals such as pesticides have widely contaminated the environment and occasionally, such as at Bhophal, India, have wreaked enormous devastation. The exploitation of oil and gas has led to numerous deadly explosions and widespread contamination through, for example, accidents to tankers such as the *Exxon Valdez*. Even energy technologies such as hydro-electric power, often thought to be benign, have led to tens of thousands of deaths from dam failures. And the ubiquitous automobile claims tens of thousands of lives every year through accidents and its emissions. We continue to use these tools because they also provide enormous benefits and they are essential to support the enormous population.

The critical question is how do we control and use not just nuclear but any advanced technology in ways that bring benefits without jeopardizing our health and the environment?

Making rational decisions requires an understanding of the science behind the technologies as well as thorough analysis and objective debate. This is not an easy task. As our society gets increasingly complex, it becomes more difficult to keep abreast. We are truly living in an age of information overload. Nevertheless, objective analysis should be the norm we strive for, rather than ill-informed responses based on emotions.

The purpose of this book is to present the science and engineering behind nuclear technology in an easy-to-understand format. We also place nuclear technology in perspective with other energy technologies. We try not to take a stand, we simply wish to remove the polemic and fears that surround this emotional topic. Our objective is to provide the information so that you can make up your own mind.

We need to progress into the new millennium in a sustainable manner — the very survival of the earth is at stake. Can we harness nuclear power to help in this struggle or do we abandon it?

Chapter Two

Splitting the Atom

THE DAWNING OF THE NUCLEAR ERA

Despite the central role that radioactivity plays in the universe and in shaping the earth, it remained effectively unknown until the late 1800s. Because radiation is invisible and emanates from the nucleus, an almost incomprehensibly tiny region that cannot be seen even with the most powerful modern microscope, it was a formidable challenge to unravel its mystery. The scientists of the late 1800s and early 1900s were ingenious indeed, given the primitive tools that were available.

Radiation was discovered accidentally by Wilhelm Roentgen in 1895 in Bavaria. Experimenting with a gas discharge tube in a darkened room, he noticed that a crystal some distance from the discharge tube would glow whenever the tube voltage was turned on, but not when the voltage was turned off. Something was being transmitted across the space between the tube and the crystal — Roentgen had discovered **X-rays.** (Words defined in the Glossary are bolded at their first occurrence and an introduction to nuclear science is given in Appendix A).

Meanwhile in France, Antoine Henri Becquerel in 1896 accidentally discovered that uranium ore produces an image on a photographic plate, although he could not explain why. Marie and Pierre Curie deciphered what was happening and gave the name radioactivity to the phenomenon. In 1903, Becquerel and the Curies jointly received the Nobel prize for their discovery of radioactivity.

Those were exhilarating days, with remarkable advances in the basic understanding of physics and chemistry. Nobel prizes, the highest recognition for scientific achievement, were awarded to many of the pioneers who advanced the knowledge of nuclear science.

In 1897, Ernest Rutherford, a New Zealander working at the famous Cavendish Laboratory at the University of Cambridge, England, discovered that radiation from uranium is composed of two parts, which he named **alpha** and **beta** particles. He also astounded his colleagues by showing that alpha or beta decay could change one **element** to a completely different one. This ran totally counter to the understanding of the atom at that time and many scientists were sceptical of this claim of alchemy. In 1911, he made another momentous discovery: he demonstrated that the atom has a nucleus that concentrates nearly all of the mass and all of the positive charge in a very small volume at the centre of the **atom**. Hans Geiger was his colleague in this famous experiment.

Although the **electron** was discovered in 1900 by J. J. Thomson at the Cavendish Laboratory, it was not until 1913 that the Danish physicist Niels Bohr proposed the model of the atom as we know it today, in which the electrons surrounding the nucleus occupy distinct energy states.

In 1932, James Chadwick, working at Cambridge University, discovered the **neutron**, which plays such a central role in nuclear fission. By that time, the broad outline was in place for our present understanding of the atom and its nucleus.

NUCLEAR FISSION

In 1905, the great physicist Albert Einstein showed theoretically that mass and energy were equivalent. It would be more than thirty years, however, before scientists discovered the immense energy that could be released by transforming matter in the **fission** process. In fact, at various times both Einstein and Rutherford despaired of ever extracting this energy. For example, in 1932 Einstein said, "There is not the slightest indication that (nuclear) energy will ever be obtainable. That would mean that the atom would have to be shattered at will."

Shortly thereafter, a Hungarian physicist, Leo Szilard, actually took out a patent on a device that would develop enormous energy from the nucleus from a **chain reaction** based on a neutron-capture

Ernest Rutherford (1871-1937), a New Zealander, is considered to be the father of nuclear physics. Although he spent most of his career in England, some of his best work was done in Canada, where he held the Chair of Physics at McGill University in Montreal from 1898 to 1907. This picture shows him in his McGill laboratory. Rutherford spent the latter part of his career at the University of Cambridge where he trained many physicists who later became prominent in Canada's nuclear development. He won the 1908 Nobel Prize for chemistry, was made a baron in 1931, and is buried in Westminster Abbey.

AECL

process involving the release of more than two neutrons. Although he had no idea whether this would work in practice, the concept was exactly how a **nuclear reactor** works.

Next came the discovery of the actual fission process itself. In 1938, two Germans, Otto Hahn and Fritz Strassman, reported the puzzling result that when they bombarded uranium with neutrons, barium and krypton were always produced. Shortly after, Lise Meitner, a Jew who had escaped from Nazi persecution in Germany to Sweden, and her nephew Otto Frisch noted that barium has 56 **protons** and krypton has 36, yielding a total of 92 protons, the same as uranium.

This clue led them to deduce that the uranium atom had been split, or undergone a process known as fission.

But there was something even more astonishing. In splitting, the uranium atom released an enormous amount of energy. In fact, the splitting of one uranium atom released seven million times the energy produced by burning one atom of carbon. The potential for creating energy from fission, including its application in weapons, was immediately recognized. The clues had all come together, including Einstein's earlier theory of mass and energy equivalence.

The expression "splitting the atom" is, of course, a technical misnomer. What is actually being "split", or fissioned, is the nucleus not the atom. Nevertheless, what should have been called "nuclear energy" was for many years called "atomic energy".

Typical fission reactions are:

$$^{235}U + n \rightarrow {}^{88}Kr + {}^{145}Ba + 3 \text{ neutrons} + \text{Energy}$$

$$^{235}U + n \rightarrow {}^{90}Sr + {}^{144}Xe + 2 \text{ neutrons} + \text{Energy}$$

Note that the total number of protons plus neutrons stays the same, that is both sides of the reaction, before and after the arrow, have the same number of nucleons. There are many different fission reactions for uranium-235. The part on the left is always the same but on the right, the fission products (e.g. the krypton-88, barium-145 for the first example reaction, and the strontium-90 and xenon-144 for the second example) can cover a large range. The most probable fission products are those with approximately one-half the mass of the original uranium atom such as strontium-90, cesium-137, iodine-129, krypton-88, xenon-133, etc. In addition, several neutrons are released (2.43 neutrons, on average, for the splitting of a uranium-235 atom); these generally have high energy, that is, high speed.

Finally, considerable energy is released. This is accompanied by a small loss of mass in the system. This is in accord with Einstein's famous equation which states that mass is a very concentrated form of energy:

$$E = mc^2$$

where E = Energy, m = mass, and c = speed of light.

THE FISSION REACTOR

What is the secret to unlocking the enormous reservoir of energy locked up in the nucleus of an uranium atom? The early scientists had noted the neutron was what caused fission of the uranium nucleus. They also noted that several neutrons were released during fission. If one of those neutrons could be made to hit another uranium nucleus, it could cause that nucleus to also fission. Then several more neutrons would be emitted that could cause more uranium atoms to fission and so on. This is called a chain reaction as depicted in Figure 2-1.

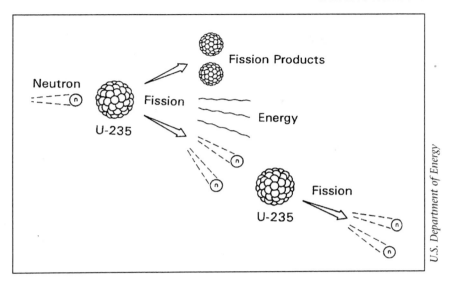

U.S. Department of Energy

Figure 2-1: A schematic diagram of a chain reaction. A neutron enters from the left and causes a U-235 nucleus to fission. This fission event produces energy, two more neutrons in this case, and fission products. The neutrons produced by the first fission then cause more fissions which in turn produce neutrons and the process proceeds in that manner.

If fewer neutrons are being generated by fission than are being used to initiate fission reactions, the process is not self-sustaining (the process is called subcritical). This is the case in nature with uranium ore bodies. If exactly the number of neutrons are being generated as are being used to split nuclei, the nuclear reaction is said to be **critical**. In this case a controlled amount of energy is being constantly released in a sustained chain reaction. This process is used in nuclear reactors. Finally, if more neutrons are being generated than are being consumed, the chain reaction can increase extremely rapidly. This process (called supercritical) is used in a nuclear bomb.

It is a complex matter to set up conditions so that the process becomes critical or supercritical. One problem is that most **nuclides** absorb neutrons and this **absorption** removes them from sustaining the chain reaction. Thus, any attempt to create a chain reaction must minimize the presence of neutron absorbers. Chain reactions do not happen in uranium ore bodies, for example, because the uranium consists mostly of uranium-238 and only has a low concentration of the fissionable uranium-235. The ore body also contains too many neutron-absorbing impurities. (There is a notable exception where a natural nuclear reaction occurred in an uranium ore body about two billion years ago; see Chapter 10.)

George Laurence (1905-1987) was born in Charlottetown, PEI, and obtained his first degree at Dalhousie University and his doctorate at the University of Cambridge as a student of Rutherford. He was the first person in the world to do research with large quantities of uranium and built a small reactor in Ottawa in 1942, which likely would have been the first in the world had materials of sufficient purity been available to him. He received many honours for his important contributions to Canadian nuclear development over a period of almost fifty years. He was President of the Atomic Energy Control Board from 1961-1970.

AECL

It took large teams of scientists many years to discover how to achieve exactly the right conditions. First, there needs to be a core of fissile material, that is, material that will fission. Uranium-235 is currently the primary material used.

By a quirk of nuclear physics, the **fissile** atom splits most readily if the bombarding neutrons are going quite slowly. As neutrons emitted by the fission process are going very fast, the core needs to be surrounded by a material called a **moderator** that slows the neutrons down. Only a few materials are good at moderating neutrons without absorbing them. The more equal the nuclear mass of the moderating material is to the mass of the neutron, the more the neutron is slowed down. It was discovered that **heavy water** (water with the hydrogen atom replaced by its isotope **deuterium**; it is about ten percent heavier than normal water) and carbon (usually in the form of graphite) are the best moderators. Ordinary water also moderates neutrons but, because of its relatively high absorption, is not as effective as heavy water.

In general, the presence of neutron-absorbing materials should be minimized in a reactor. They can, however, be used to stop or control the nuclear fission process. For example, neutron-absorbing control rods can be moved into and out of the core to control the reaction. Substances that are particularly good at absorbing neutrons and, thus, "poisoning" the fission reaction are boron, cadmium, and gadolinium.

The engineering behind nuclear reactors is described in Chapters 6 and 7.

THE RACE FOR NUCLEAR WEAPONS

The dark shadow of World War II now falls across the nuclear story. Spurred by the threat of their antagonists gaining the bomb, scientists and engineers on both sides of the conflict worked feverishly, compressing what would likely have taken many decades into a few years. Working in Italy, Enrico Fermi showed that when neutrons are slowed down by moderators such as graphite and water they could be absorbed by nuclei. He applied his theory to make several radioactive materials, for which he was awarded the Nobel prize in 1938. He used the trip to the award ceremony in Stockholm to escape from fascist Italy, and later played a key role in the US nuclear program.

On the eve of the Second World War, Einstein wrote to US President Franklin D. Roosevelt, warning him that Nazi Germany was working to develop the ultimate weapon of destruction. This led President Roosevelt to establish a committee to direct research on this vital topic and in 1942, shrouded in secrecy, the famous Manhattan Project was initiated.

By the fall of that year, under Fermi's direction, the world's first nuclear reactor was constructed at the University of Chicago. This primitive reactor was known as a "pile" because it consisted of a pile of graphite blocks and natural uranium into which were inserted neutron-absorbing rods to control the reaction. The first man-made self-sustaining fission chain reaction took place on December 2, 1942. Uranium, used for adding colour to glass and ceramics for hundreds of years, now took centre stage as a powerful new energy source.

Other important nuclear laboratories were established at Oak Ridge, Tennessee, Hanford, Washington, and Los Alamos, New Mexico, culminating in the development of the atomic bombs that were exploded over Japan. Those bombs were to have a profound effect on the outcome of the war as well as on post-war developments. Indeed, those bombs still cast an ominous shadow on modern society.

CANADA STARTS ITS NUCLEAR PROGRAM

The nuclear era in Canada began in the midst of the frantic race between fascist Germany and the free world to prepare an atom bomb. The Canadian team included French and British scientists who had been evacuated from Europe, of which Sir John Cockcroft, who later became Director of the Canadian effort, was one of the more eminent members. Cockcroft was a student of Rutherford and his research from the Cavendish Laboratory in Cambridge, England, was transferred to Canada.

Those were drama-filled years. Two scientists, Hans Von Halban and Lew Kowarski, escaped from the Institut du Radium in Paris only steps ahead of the invading German army. They took with them 200 kilograms of heavy water the world's total supply that the Germans were trying desperately to find. Both the scientists and the heavy water became key parts of the embryonic nuclear research effort in Canada.

Canada's program was designed to support the larger American project by developing a heavy-water reactor to produce plutonium, a prime fissile ingredient for the atomic bomb. Initially located in Montreal, in 1944 the project team moved to specially constructed laboratories at Chalk River, Ontario, 200 kilometres northwest of Ottawa on the Ottawa River. The research program was directed by the National Research Council which was already involved in nuclear research.

Uranium, up to then an unwanted byproduct of the Eldorado radium refinery at Port Hope, suddenly became a very valuable commodity, as did the Eldorado refinery, the only radium refinery in North America.

The objective of Canada's nuclear program was to construct a reactor moderated by heavy water with thermal power output of 20 to 40 **megawatts** (million watts of power). To better understand the physics and material problems, a small prototype reactor named **ZEEP** (Zero Energy Experimental Pile) was designed initially and went critical on September 5, 1945. Figure 2-2 shows the ZEEP reactor. This was a remarkable achievement, as Canada became only the second country in the world, after the USA, to control nuclear fission using a reactor. ZEEP was permanently shut down in 1970.

In October 1946, the Atomic Energy Control Board was created in Canada to promote the development of peaceful applications of nuclear technology. [The Atomic Energy Control Board (**AECB**) became the Canadian Nuclear Safety Commission (CNSC) in 2000. To avoid confusion, the **CNSC** will be used even for events that occurred when it was called the AECB.] Canada has an exemplary track record in this regard. Although one of the world's leading countries in nuclear

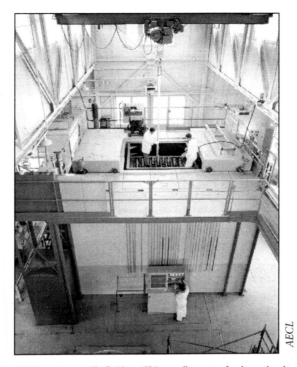

AECL

Figure 2-2: The ZEEP reactor at Chalk River. This small reactor for investigating the physics of fission was the first nuclear reactor built outside the USA, starting up in 1945 and operating until 1970.

technology, and having the capability to develop nuclear weapons since 1946, Canada has chosen not to do so. In fact, a weapons program for Canada was never seriously considered, something all Canadians can take pride in.

Following the War, Canada continued to support the US and British weapons programs as a member of the NATO alliance, and continued its efforts to develop a powerful research reactor. In 1947, the 20 MW thermal **NRX** reactor (National Reactor Experimental) came on line at Chalk River. Although its original purpose was to supply weapons-grade plutonium for the US, mechanisms for producing isotopes and channels for conducting experiments with beams of neutrons were also built into the reactor. NRX generated a very high neutron flux (number of neutrons per second per unit area) and for many years was the world's most powerful and best research reactor, while at the same time producing plutonium. In 1965, under Lester Pearson's government, Canada made the policy decision to cease contributing to the US weapons program.

THERMAL AND ELECTRICAL ENERGY

The power of reactors is measured in the unit megawatt (MW) or one million watts. For research reactors, which do not generate electricity, the power is quoted in thermal megawatts. For example, the McMaster University Reactor is rated at 5 MW (thermal) or 5 MWt.

For a power reactor, electricity, not heat, is the quantity of interest. Even with modern equipment, only about 33 percent of the heat created by a nuclear reactor is converted to electricity. Thus, for a CANDU reactor of the type located at Pickering, capacity is quoted as 515 MW, that is, it can generate 515 MW of electricity (even though it generates about 1,550 MW of thermal power). This can also be written as 515 MW(electrical) or 515 MWe.

In this book, MWt always means thermal megawatts, and MW always means electrical megawatts.

Following an accident in 1952 (see Chapter 8), NRX was upgraded to 40 MWt. NRX (see Figure 2-3) had a long history of producing radionuclides for medical applications that began in 1949 with sales of iodine-131, phosphorus-32, carbon-14, cobalt-60 and others. NRX has also had an illustrious career in research which is discussed further in Chapter 16. NRX was finally taken out of service in 1992, at the ripe age of 45 — a remarkable feat considering that its useful life was originally estimated to be about five years.

In 1952, Atomic Energy of Canada Limited (**AECL**), a Crown Corporation, was formed and took over responsibility for Canada's nuclear program from the National Research Council.

In 1957, a larger reactor, the **NRU** (National Research Universal) was built in Chalk River. At 135 MWt, the NRU is a world-class facility for conducting research and producing **isotopes** for medical and industrial uses.

Although neither the NRX nor NRU reactors were used to generate electricity, they were the beginning of the **CANDU** (Canada Deuterium Uranium) power-reactor concept. In particular, the NRU is both heavy-water moderated and cooled, and also features on-power refuelling (that is, fresh fuel can be inserted and used fuel removed without shutting down the reactor). Unlike the future CANDU reactor with a horizontal core, the NRU reactor core is vertical and fuel is loaded from the top using a transfer flask. NRU is still operating.

These two reactors have been veritable workhorses and are responsible for a wealth of scientific research as well as for the production of isotopes that have been used in medical and industrial facilities throughout the world (see Chapters 11 and 12). The availability of these isotopes stimulated the development of medical treatment systems. Figure 2-4 shows one of the earliest cobalt therapy units for cancer treatment.

Figure 2-3: The NRX research reactor at Chalk River first saw operation in 1947 and for many years was a world leader in nuclear research because of the large number of neutrons it produced and its excellent experimental instrumentation. In the foreground a scientist can be seen at one of the neutron spectrometers. The reactor was shut down in 1992.

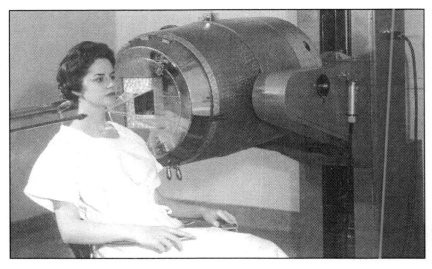

AECL

Figure 2-4: The first cobalt-60 cancer therapy machine, built in 1951. Since then thousands of similar units have been manufactured in Canada and used to treat millions of patients throughout the world.

THE CANDU NUCLEAR POWER REACTOR

By the early fifties the idea of using nuclear fission to produce electrical power was becoming an achievable goal. Based on the Chalk River developments, AECL, Ontario's electrical utility, Ontario Power Generation (then named Ontario Hydro), and General Electric Canada joined forces to design and construct the Nuclear Power Demonstration (**NPD**) reactor. [Ontario Power Generation (**OPG**) was formed from Ontario Hydro in 2000. To avoid confusion we will use OPG throughout this book even for those times when the company was called Ontario Hydro.] Located at Rolphton on the Ottawa River just a few kilometres upstream from the Chalk River site, the 22 MW reactor supplied its first power to the Ontario grid in 1962.

NPD set the stage for the unique CANDU design that was to become a Canadian trademark: NPD was fuelled by natural uranium, moderated and cooled by heavy water, used thin (4 mm) Zircaloy (an alloy of zirconium containing about 1.5% tin) fuel channels instead of a thick (15-20 mm) steel pressure vessel typical of other reactor types, and could be refuelled while operating. The NPD station became the training centre for nuclear staff and generated electricity until 1987, paving the way for the ambitious nuclear program subsequently implemented by Ontario Power Generation.

From that small beginning CANDU reactors can now be found producing electricity in Ontario, Quebec, New Brunswick, Argentina, India, Pakistan, Korea, Romania, and, beginning in 2002, China.

The next step in developing a commercial-scale reactor was the Douglas Point reactor located on the east shore of Lake Huron, north of Kincardine. The decision to start was made in 1959, and AECL and Ontario Power Generation collaborated in the design and construction. Whereas NPD was strictly a pilot project, Douglas Point was a prototype power reactor whose purpose was to show that the small NPD reactor could be scaled up to commercial size and operated safely and economically. The power output of the Douglas Point plant was an order of magnitude larger at 206 MW, and placed Canada at the forefront of nuclear engineering alongside the USA with its Shippingport reactor and the UK with its 300 MW Magnox reactor.

Douglas Point went critical November 15, 1966, the first electricity was delivered in 1967, and it was declared in-service in September 1968. The cost was $91 million. Owned by AECL, Douglas Point was operated by Ontario Power Generation until May 5, 1984.

Many lessons were learned from Douglas Point. The on-line fuelling machines proved to be troublesome and required a number of changes; there were losses of heavy water until inadequate valves and flanges were redesigned; a proprietary process for decontaminating moderator and coolant was developed; a remotely controlled "mouse" was developed that crawled into the bowels of the reactor and fixed a leak in 1976. The annual capacity factor (the percentage of time that a reactor is available for generating electricity) grew from 54% at outset to 82% on retirement. In its later years, Douglas Point

Table 2.1
HISTORICAL DEVELOPMENT
OF NUCLEAR POWER IN CANADA

Era	Reactor	Power	Location
1945-1970	Zero Energy Experimental Pile (ZEEP)	0	Chalk River
1947-1992	National Research Experimental (NRX)	42 MWt	Chalk River
1957-now	National Research Universal (NRU)	135 MWt	Chalk River
1962-1987	Nuclear Power Demonstration (NPD)	22 MW	Rolphton
1968-1984	Douglas Point	206 MW	Douglas Point

also supplied steam for a heavy-water plant located nearby. Douglas Point proved to be a difficult plant to operate and would have required extensive refits to keep it running economically. In 1984, with the four Pickering A stations and four Bruce A stations already operating, and Pickering B and Bruce B just starting to come on line, the technical effort to operate Douglas Point could not be justified, and the reactor was shut permanently. The lessons learned at Douglas Point contributed enormously to these larger stations being successfully constructed and operated.

Ontario Power Generation was so confident with the progress made with NPD and Douglas Point that it committed to the construction of four reactors at Pickering, Ontario, before Douglas Point even came into service. The four reactors that comprise station A came on line from 1971 to 1973 with a capacity of 515 MW each (Figure 2-5). The Pickering site consists of 271 hectares and is about 32 kilometres east of Toronto on the shore of Lake Ontario. From 1983 to 1985 station B, consisting of four more reactors, came into operation at Pickering. These reactors have a capacity of 508 MW each and were constructed adjacent to Station A so the eight reactors are in a long row. The eight reactors share one large vacuum building, a safety device that limits the consequences of an accident (see Chapter 8). When all eight reactors are in operation, the station generates 4,092 MW of electricity. This is an enormous amount of electricity, approximately double what Ontario Power Generation generates at Niagara Falls, one of the largest hydroelectric stations in the world.

The Pickering site includes a visitor's centre, a 20-hectare wildlife sanctuary and recreation area. A fish farm operated until 1998 producing yellow perch, trout, and American eels using warm water from the Pickering cooling system.

During the 1970s and 1980s, Ontario's appetite for electrical power was voracious. With almost all potential hydro sites developed, the province turned to nuclear to supply the power it needed to keep industry and commerce humming. Ontario became the industrial heartland of Canada, thanks in part to the supply of abundant nuclear electricity.

The capacity of the reactors at Pickering was not sufficient to meet the continuing growth in electrical demand and major reactor construction was commenced at the Bruce Nuclear Power Development site, located on Lake Huron about 255 kilometres

Ontario Power Generation

Figure 2-5: The Pickering Nuclear Generating Station. This station on Lake Ontario east of Toronto consists of 8 reactors in two groups of four. The reactors have the rounded domes. The larger cylindrical structure is the vacuum building designed to capture and quench any radioactive steam from an accident.

northwest of Toronto. From 1977 to 1979, the Bruce A station, consisting of four reactors, came on-line. Each reactor supplied 750 MW of electricity as well as producing steam for heavy water plants and the Bruce Energy Centre. From 1984 to 1987, the Bruce B station came into operation with four reactors, each having 850 MW capacity. Unlike Pickering, the two Bruce stations are geographically separated. When all eight reactors are operating, the Bruce site can deliver up to 6,400 MW of electricity to Ontario's grid, a prodigious amount of energy. All of the reactors are of CANDU design and are owned and operated by Ontario Power Generation.

In 2001, following a policy decision by the Ontario government to privatize the province's electricity industry, Bruce Power took over operation of the Bruce reactors under a long-term lease agreement (18 years plus a 25 year option) with Ontario Power Generation. Bruce Power is owned by British Energy (85%) and Cameco Corporation (15%), Canada's largest uranium supplier (see Chapter 13). By 2003, the Power Workers Union and the Society of Energy Professionals will acquire up to 5.2% ownership.

With eight reactors, the Bruce site, which occupies 920 hectares, is one of the largest nuclear centres in the world and was known as the "Nuclear Capital of the World". It also contains a heavy-water plant (closed in 1998), a heavy-water plant that was mothballed in 1984, and two that were never completed. The heavy water was produced by exchanging deuterium between water molecules and hydrogen sulphide (H_2S) gas. The process is based on the fact that at certain temperatures a deuterium atom prefers to become part of an H_2S molecule rather than remain in a water molecule. The problem with these plants was that H_2S is a highly poisonous gas and it is doubtful that new plants using this process would be acceptable under today's stringent environment assessment processes. The Bruce A reactors supplied process steam to the heavy water plant. The site also contains a waste management area where low and intermediate-level wastes from all of Ontario Power Generation's facilities are treated and stored. The Douglas Point reactor, now decommissioned, is also on site. Adjacent to the nuclear park is the Bruce Energy Centre, an industrial park where greenhouses and factories utilized process steam from the Bruce A reactors, prior to their shutdown. Steam is currently being supplied by oil-fired boilers until the Bruce A reactors are brought back into service.

Meanwhile, Quebec, although possessing much larger hydroelectric potential than Ontario, also made the decision to enter the nuclear age. In 1971, the 250 MW Gentilly 1 reactor came into operation at Gentilly near Trois Rivières on the shores of the St. Lawrence River. Built by AECL, the reactor was known as the CANDU-BLW, a type of CANDU reactor moderated by heavy water but cooled by boiling light water, rather than by pressurized heavy water. The reactor had a vertical pressure-tube core unlike all other CANDU designs, which are horizontal in orientation. Gentilly 1 was a back-up design to the standard CANDU design because AECL was concerned that heavy-water losses at its other nuclear stations might prove to be unacceptably high. Gentilly 1 was plagued by problems from the outset and was commonly referred to in Quebec to as "le citron (the lemon)." Unable to compete economically with other sources of electricity in Quebec, it was taken out of service in 1977. The reactor was used for training until it was mothballed in 1979. Heavy-water losses at Douglas Point and Pickering proved to be quite low and further development of the CANDU-BLW reactor was curtailed. However, Gentilly 2, a 685 MW CANDU 6 reactor was constructed at the same site and started to deliver electricity to Hydro Quebec's grid in 1983. It has operated successfully and is expected to reach its 30-year service life in 2013.

New Brunswick also produces nuclear power. A 634 MW reactor of the CANDU 6 type (see Figure 2-6) was commissioned at Point Lepreau in 1982 and supplies about a third of the province's electrical needs. Whereas Ontario Power Generation's reactors were all designed in multi-unit stations, the CANDU 6 is a single, stand-alone reactor designed and marketed by AECL.

New Brunswick Power

Figure 2-6: The Point Lepreau nuclear station operated by New Brunswick Power consists of a single reactor of the CANDU 6 type.

Four reactors of 880 MW each were constructed at Darlington about 70 kilometres east of Toronto on the shores of Lake Ontario. Begun in 1977, they were delayed for about four and a half years, primarily because Ontario's electricity demand finally slowed down. This led to a considerable increase in the cost of these reactors due to the very high interest rates during that period. The four reactors were connected to Ontario's electrical grid between 1989 and 1992 and supply up to 3,520 MW of electricity.

The 485-hectare Darlington site also houses the Tritium Removal Facility, only the second in the world, which extracts **tritium** from heavy water that has been used in Ontario Power Generation's reactors. Tritium is formed by neutron irradiation of the heavy water used as moderator and coolant inside a reactor's core. About 2.5 kilograms of tritium are extracted each year from the heavy water to decrease the radiation fields to which station workers are exposed. Tritium is used to illuminate Exit signs and emergency markers so that they can be seen even when electrical power is interrupted. Biomedical research employs tritium as a tracer to track metabolism or movements of drugs and other substances in the body. Tritium is also used in fusion energy research (see Chapter 15).

Those were exciting years for the Canadian nuclear industry with 23 power reactors constructed in the two decades spanning the 1970s and 1980s, more than one per year. In 1992, with the completion of Darlington, there were 22 CANDU reactors operating in the country with a total capacity of 15,200 MW. Twenty of these were in Ontario supplying about two-thirds of the province's electrical power (13,590 MW). Nuclear power supplied 19% of Canada's electricity in 1994. Table 2-2 summarizes the CANDU reactors in Canada.

Table 2-2
SUMMARY OF CANDU REACTORS IN CANADA

Name	No. Units	MW/unit	Started	Closed
Douglas Point	1	206	1968	1984
Pickering A	4	515	1971-73	
Pickering B	4	508	1983-86	
Bruce A	4	750	1977-79	
Bruce B	4	850	1984-87	
Darlington	4	880	1989-92	
Gentilly 1	1	250	1971	1977
Gentilly 2	1	685	1983	
Point Lepreau	1	634	1982	

In the latter half of the 1990s, nuclear power was to suffer badly. A dramatic decrease in the growth of electricity demand, for the first time that century, cancelled plans for additional nuclear capacity. In addition, personnel cutbacks and neglect of proper maintenance by Ontario Power Generation caused significant deterioration in some of their units. This led to extended shutdowns of the Bruce A and Pickering A reactors in 1998 to develop improved reactor management practices (see Chapter 6 for further details).

Chapter Three

Radiation Everywhere

For some, nuclear technology and the science behind it are difficult to comprehend and, therefore, are frightening. But nuclear technology is not alone in this regard. We live in a technological age. Our modern world is full of scientific marvels that are difficult to understand, and the boundaries of science and technology are constantly being expanded. Today's scientists have travelled far beyond the nucleus. Now they are unravelling particles that are even smaller and more mysterious, such as leptons, muons, pions, and quarks. And who can comprehend how tiny, barely visible slivers of semiconductor chips can store millions of bits of data or compute millions of calculations per second? And there are the marvels of telecommunication, satellites and space exploration, astronomical telescopes that peer light-years into distant galaxies, organ transplants in medicine, and much, much more. It is difficult for any individual to keep abreast of all the new developments.

For those who are frightened of nuclear technology and its complexity, it may be reassuring to know that radioactivity is something natural. Many people do not realize it, but radiation is everywhere around us. Radioactivity is a natural and integral part of the earth. It is as common — and necessary — as the oxygen we breathe and the sunlight that brings life to our planet.

Before we look at nuclear technology in more detail, let us explore the role radiation plays in the natural world.

TERRESTRIAL RADIATION

Radioactive elements are found in all the rocks of which our earth is composed and have been present since the earth was formed 4.5 billion years ago. In fact, these elements were formed during the creation of the universe in a complex process called nucleosynthesis, which involved building up the various elements from primordial matter.

When we gaze in awe at the snow-capped peaks of the mighty Rocky Mountains, radioactivity does not usually spring to mind. Yet it is natural radioactivity that is responsible for the formation of the Rocky Mountains, the Alps, the Himalayas, and all mountain chains. The earth's interior would have cooled down long ago if it were not for the heat generated by the decay of radioactive elements, primarily uranium, thorium and potassium, found in rocks. Radioactivity is the energy source that drives continental drift and **plate tectonics.** As the continent-sized plates move about on the earth's surface, mountain chains form where the plates bump into each other. Thus, radioactivity-driven continental drift is constantly generating mountains.

Tourism Yukon

Figure 3-1: The St. Elias mountains in the Yukon Territory. Mountains are formed by the collision of tectonic plates, which are driven by radioactive elements in the earth.

Without this on-going process, mountains would long ago have been eroded by the forces of wind, rain, and snow, and the earth would be a smooth, featureless globe totally covered by water. Needless to say, life would have evolved completely differently. It is interesting to speculate what kind of fish form we humans might be today if there were no radioactivity on earth.

Because radioactivity is constantly decaying away, the earth and our environment were more radioactive in the past. For example, of the uranium-235 present when the earth formed, only about 1% remains today. Although greatly diminished, radioactivity continues to shape the earth we live on, and is an important tool that geoscientists use to date the age of different rocks and to decipher the history of the earth.

The radioactivity from rocks enters soils, building materials, and virtually everything on earth. Radioactivity enters human bodies when we eat plants, which take up radioactivity through their roots. Soils also emit radon-222, a radioactive gas which arises from the uranium-238 decay chain (see Appendix A), and radon gas becomes a component of the air we breathe. Thus, all rocks, all soils, all plants, and all living things contain natural radioactivity. In fact, it is impossible to find or construct a place on earth that is completely free of radiation.

Potassium-40, thorium-232, and uranium-238, known as **primordial radionuclides** because they were created at the beginning of the universe, are the most significant sources of radiation in the environment because of their radioactive decay modes, **abundances,** and **half-lives.** As discussed in Appendix A, each of uranium-235, uranium-238, and thorium-232 decays through a long series of radioactive **daughters** (decay products) until a stable isotope of lead is reached.

Neither uranium nor thorium themselves are particularly hazardous for human health because neither participates to any significant degree in plant or animal metabolism. Some of their daughters, however, are important, particularly radium-226 and radium-228, because they bear a chemical similarity to calcium.

Table 3-1 shows six of the 18 radionuclides that occur on earth but do not arise from the uranium or thorium decay chains. These have very long half-lives and many are exotic elements with high mass numbers. Most of the elements occur as mixtures of isotopes. For example, potassium has nine isotopes, ranging from potassium-37 to potassium-45. Six of these isotopes decay rapidly and are not found in nature. Two of them are stable (potassium-39 and potassium-41) and together with the small percentage of radioactive potassium-40 make up naturally occurring potassium. Rubidium-87, for example, has almost the same abundance as potassium-40 but is of minor interest because the beta radiation it emits as it decays to strontium-87 is very weak. In contrast, potassium-40 emits a very energetic **gamma ray.**

Table 3-1: SOME PRIMORDIAL TERRESTRIAL RADIONUCLIDES

Isotope	Half-life
Potassium-40	1.27×10^9 years
Vanadium-50	$>4 \times 10^{16}$ years
Rubidium-8	74.8×10^{10} years
Indium-11	31.2×10^{13} years
Lanthanum-13	81.1×10^{11} years

Table 3-2 shows typical naturally-occurring uranium concentrations. It is seen that the oceans are characterized by low concentrations of uranium (and the other main radioactive nuclides). When our distant ancestors emerged from the sea to crawl and eventually to walk on land, they entered an environment with hundreds of times more radiation.

The amount of uranium and thorium in different rock types varies considerably from place to place depending on the local geology. In some areas the radioactive elements have been concentrated by natural processes, sometimes to the point where they can be economically mined. Uranium ore bodies are found in Canada in northern Saskatchewan (see Chapter 13) as well as in Australia, Russia, the USA, and other countries. The concentration of uranium in these ore bodies, which are found from near the surface to several kilometres deep, ranges from about 0.1% to over 15%. These ore bodies provide considerable information on how radioactive materials behave in our environment and the impact that they have on human health.

Uranium and thorium are contained in the minerals zircon and monazite which are dense and highly resistant to chemical weathering. Prolonged wave action on beaches is particularly effective in winnowing out less dense quartz and other minerals, leaving the more

Table 3-2 TYPICAL NATURALLY OCCURRING URANIUM CONCENTRATIONS
(Parts per million Uranium)

High-grade ore body (10% Uranium)	100,000
Low grade ore body (0.1% Uranium)	1,000
Granite	4
Sedimentary rock	2
Average in continental crust	1.4
Seawater	0.003

radioactive elements. Significant concentrations of these radioactive minerals are found in Brazil and southwest India. The latter contains the highest thorium concentrations (8-10%) in the world.

Other areas with high terrestrial radioactivity include western Australia, southwest England, and southeastern USA. High radium levels are found in the water of Illinois and Iowa. The most remarkable instance of high natural radiation levels is in Ramsar, Iran where there are up to 50 hot springs delivering large amounts of radium into the environment. The inhabitants of these places are exposed to considerably more radiation than people living elsewhere.

In summary, natural radiation is quite variable depending largely on the local geology. People living in areas of limestone receive lower **doses** of radiation than those living in granitic areas. In extreme cases, such as in one area of Brazil where people live on soils containing abnormally high concentrations of thorium, the external radiation dose can be up to 120 times greater than found in normal areas. As explained in Chapter 4, people living in such high-natural-radiation areas experience no adverse health effects.

RADON

Although radon is a form of terrestrial radiation, it merits separate discussion because it is the largest component (approximately half) of the natural radiation to which we are exposed. Radon is a colourless and odourless gas formed from the radioactive decay of radium, which in turn is one of the daughter products of the uranium decay chain. In the uranium-238 series, radium-226 decays to radon-222, which has a half-life of 3.8 days (see Table A-2). Radon is also produced in the thorium-232 series, where radium-228 decays, via some intermediate daughters, to radon-220. Since this has a half-life of only 53 seconds, it generally decays before escaping from the soil and is not as significant a concern as radon-222.

Radon is constantly being produced and released from the ground and is always present in air. Outdoors it is greatly diluted by fresh air and is generally in low concentrations that are not a health concern. It can, however, seep into and collect in buildings and become more concentrated. As illustrated in Figure 3-2, seepage into a house takes place through cracks, joints, dirt floors, and around drainage pipes and sump pumps. Indoor levels of radon can vary greatly depending on the location of a building and how it is sealed and ventilated.

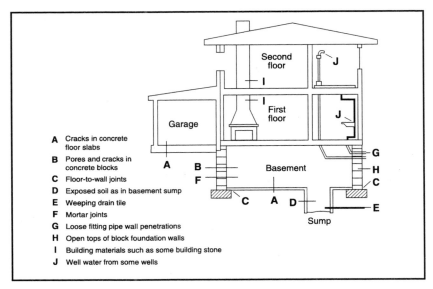

Figure 3-2: Schematic cross section of a house showing entry points for radon. While steps can be taken to block radon entry at these points, the key to radon control is good ventilation.

It is not the radon itself that is a health concern, but the radioactive daughter products into which it decays. Radon, being a gas, is simply the vehicle by which members of the uranium decay chain can escape from the soil and enter our lungs. Outside the body, radon is not a concern since the alpha particles it emits cannot penetrate the skin.

Furthermore, being chemically inert (it is one of the noble gases), radon can be inhaled and exhaled harmlessly. However, radon decays into daughter products that are solid and, having an electrical charge, can attach themselves to dust particles. In this form, the radon daughters can lodge in the bronchial and lung tissues where they emit alpha particles that can damage surrounding cells and, if concentrations are high enough, could cause cancer.

Radon is the subject of considerable research and debate. The US Centers for Disease Control and Prevention believe radon to be the second greatest cause of lung cancer, next to smoking, and estimates it may cause between 5 and 10% of all lung cancers. Because of this concern, the Surgeon General of the United States urged that most homes be tested for radon levels. However, a recent study by Cohen (1995) that included 90% of the population of the United States showed that those counties with higher radon concentrations had lower lung-cancer mortalities.

In Canada, Health Canada has not recommended testing of homes although it has established a guideline of 800 **becquerels**/cubic metre (see Appendix A for an explanation of these units) for the maximum acceptable level of radon. This is five times higher than the US guideline and four times greater than that of Great Britain. Based on measurements across Canada, it is estimated that less than one home in a thousand requires corrective action.

The radon level in a home can be measured by activated-charcoal detectors. These consist of containers that allow air to enter, and then radon and its daughters are absorbed by the charcoal. These detectors are generally exposed for a day or two and then sent to a laboratory for analysis. They are easy to use and inexpensive. Due to the brief sampling period, they are best used for screening to determine if more detailed studies are needed.

Alpha track detectors, which use a sheet of polycarbonate plastic in a container, are also easy to use. The radioactive emissions of alpha particles cause small damage marks in the plastic which are subsequently measured with suitable instruments. The container is generally left in place for at least three months.

Radon levels in homes can be lowered relatively simply and inexpensively by improving circulation and ventilation and by sealing cracks in the foundation and around pipes and sump pumps.

COSMIC RADIATION

In 1911, the British scientist C. T. Wilson invented a device now called the **Wilson Cloud Chamber**, which allowed radiation to be seen. A cloud chamber experiment is so simple that it can be done in a high-school laboratory or even at home. A clear glass or plastic container forms the chamber, and ethyl alcohol, super cooled by dry ice, is used to form a "cloud". As radiation passes through the chamber, it knocks electrons off (**ionizes**) the atoms in the air. The alcohol vapour then condenses on the charged particles, forming tracks similar to the vapour trails of jet planes (See Figure 3-3). These are easily visible by shining a light through the chamber in a darkened room. The tracks disappear relatively quickly. Most of the tracks will be about one centimetre long and are quite distinct; these are caused by alpha particles. Beta particles cause longer, thinner tracks. Occasionally, faint, twisting, circling tracks are observed; these are due to gamma rays.

The cloud chamber visually confirmed the momentous discovery that the earth is being bombarded by intense radiation originating from space. Although it took decades to understand the processes involved, it was soon realized that nuclear reactions drive the universe, providing the fundamental power to make stars shine and give light and warmth to planets like the earth. All stars, including our sun, are giant nuclear reactors powered by a process called nuclear **fusion** (described in Chapter 15). Not only does the sun create the light and heat on which our world depends, but the giant inferno inside the sun is constantly ejecting a stream of energy and particles, called the **solar wind**, into space. The amount of this cosmic radiation increases considerably during solar flares.

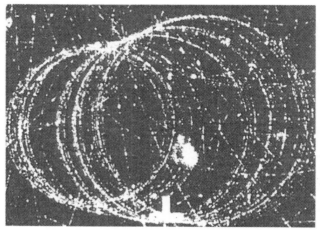

Figure 3-3: Cloud chamber showing vapour trails caused by cosmic radiation. Some trails are circular because a magnetic field was applied in this photograph to help in identifying the masses of the particles.

The particles in the solar wind travel through space and react with the earth's atmosphere creating cosmic radiation that constantly rains down on us. The solar wind consists largely of protons together with alpha particles, a variety of heavier particles, electrons, **photons**, and **neutrinos**. Although most of these particles come from our sun, some originate in far-off suns and galaxies. The solar wind interacts with the upper atmosphere to ionize some gases and also to create new (**cosmogenic**) radionuclides. This radiation extends from the upper atmosphere all the way to the earth's surface. Thus, all living beings are constantly bombarded by millions of particles of cosmic nuclear radiation each second, whether they live on earth or on planets in other corners of the vast universe. Table 3-3 lists some of the radionuclides that have a cosmic origin.

Table 3-3
RADIONUCLIDES OF COSMIC ORIGIN

Isotope	Half-life	Production Rate*
Hydrogen-3 (tritium)	12.3 years	0.25
Beryllium-7	53.6 days	8.1×10^{-3}
Beryllium-10	2.5×10^6 years	3.6×10^{-2}
Carbon-14	5730 years	2.2
Sodium-22	2.6 years	5.6×10^{-5}
Sodium-24	15 hours	**
Silicon-32	650 years	1.6×10^{-4}
Phosphorous-32	14.3 days	8.1×10^{-4}
Phosphorous-33	24.4 days	6.8×10^{-4}
Sulphur-35	88 days	1.4×10^{-3}
Sulphur-38	2.87 hours	**
Chlorine-36	3.1×10^5 years	1.1×10^{-3}
Chlorine-38	37 minutes	**
Chlorine-39	55 minutes	1.6×10^{-3}

* in atoms per square centimetre per second
**half life too small to measure production rate

The amount of cosmic radiation is least at sea level because of the shielding provided by the atmosphere and increases progressively as the height above sea level increases. Residents of Banff, for example, receive 0.2 milli**sieverts** (mSv) per year (see Appendix A for an explanation of mSv) more radiation than the inhabitants of Halifax. Flights in aeroplanes yield more radiation than staying on the ground. For example, a cross-country aeroplane flight yields an additional 0.02 to 0.03 mSv of cosmic radiation than if one had stayed at ground level (perhaps a small negative side to being a jet setter). The radiation dose would be even greater for travel by Concorde, as it flies at a substantially higher altitude (18 kilometres) than commercial air liners (11 kilometres). Concordes are equipped with radiation monitors so that if a solar flare erupts (when radiation can increase by up to 10,000 times), they can descend to lower altitudes. Commercial aircraft crew receive about 1 to 2 mSv/yr of radiation from flying; Concorde crew members receive about 2.5 mSv/yr from flying (about the amount of radiation the average Canadian is exposed to in one year).

Astronauts who travel in space on lunar visits or to orbit around the earth receive even more radiation, particularly if they stay "up" for an extended period of time. On the Apollo XIV mission to the moon in January 1971, it was estimated that the astronauts received a radiation dose of approximately 5 mSv in 12 days (about two annual background doses). This was a real concern when the NASA space program first started, and considerable research has been done to study the effect of cosmic radiation not only on astronauts but also on electronics and other delicate control systems used in space.

A particular concern continues to be the huge increase in cosmic radiation that occurs when the sun has solar flares. Space missions are scheduled to avoid solar flares whenever possible. If humans are ever to travel to distant solar systems, as portrayed in science-fiction books, they will be exposed to elevated levels of cosmic radiation for many years during their intergalactic journey.

RADIATION INSIDE THE BODY

Since there is so much radiation in the environment it is not surprising that there is substantial radiation in our bodies. Potassium-40 is the most important radioactive component of food and human tissue. About 120 out of every million atoms of potassium (0.012%) are radioactive; the rest are stable. Potassium is a component of soil and from there is taken up by plant roots. It comes into our bodies directly when we eat fruit and vegetables and indirectly via the meat of animals who eat root crops. The radioactive potassium is then deposited in parts of our body such as bones. Potassium helps maintain fluid pressure and balance within cells. It is also important in normal muscle and nerve response and in maintaining heart rhythm. Even a temporary potassium deficiency can result in serious upsets of body functions.

Scientists have established that the potassium-40 in our bodies undergoes about 60 decays per second per kilogram of body weight. Thus, approximately 4,500 radioactive decays per second, each emitting a gamma rays, happen inside a person weighing 75 kilograms (165 pounds). Combined with other natural radioisotopes inside and outside our bodies, this person is struck by nuclear radiation about 54 million times in a single hour. Every day of our existence over a billion radioactive particles pass through our body. Through the long evolution of humans, our bodies have learned how to live with this radioactivity.

Carbon-14 is another important biological element, since carbon forms about 23% of our bodies by weight. Carbon is a major component of the food we eat, forming about 45% of carbohydrates, about 55% of fat, and about 50% of proteins. Carbon-14 is responsible for almost as much radioactivity within our bodies as potassium-40 (about 3,900 decays per second for a person weighing 75 kilograms).

Radioactivity is not only a natural part of our environment, but also of our bodies.

HUMAN-MADE RADIATION

Not all radiation in the environment is natural; some arises from human activities. The largest human-made source of radiation is from medical applications. Other very small contributors are nuclear laboratories, industrial and consumer sources such as smoke detectors, and atmospheric fallout from nuclear weapons testing. In addition, nuclear power reactors contribute about 0.3% of the radiation in the environment.

About 90% of the medical radiation dose comes from X-rays, with each person in Canada receiving, on average, 1.2 X-rays per year. Of these, 14% are dental X-rays. Other medical radiation doses come from radioactive isotopes used in various diagnostic tracer tests involving small doses.

Technically, the radiation from X-rays does not emanate from the nucleus, but is caused by transitions in the innermost shell of electrons. Thus, X-rays are an electromagnetic, not a nuclear, phenomenon. The ionizing effects, however, are the same as those caused by nuclear radiation.

Coal-fired power plants release radiation in their emissions due to the radioactive elements in coal. Radiation can arise from the release of radon from disturbing the earth during construction and road-building projects, and from the use of phosphate fertilizers, which contain relatively high concentrations of natural radioactive elements.

Consumer products that emit very small amounts of radiation include colour televisions and smoke detectors.

From 1945 to 1963, the Soviet Union, the USA, Great Britain and France developed nuclear weapons, and tested them by detonating them above ground. These tests sent radioactive materials high into the atmosphere where they were dispersed by winds and carried around the globe. The radionuclides of principal biological concern were tritium (12-year half-life), strontium-90 (28 years), cesium-137

(30 years), iodine-131 (8 days), manganese-54 (314 days), iron-55 (2.7 years), and cobalt-60 (5.3 years). By the late 1950s, the increase in atmospheric radiation was measurable and had become a concern.

On August 5, 1963, the United States, the Soviet Union, and Great Britain signed the *Treaty Banning Weapons Testing in The Atmosphere, in Outer Space, and Under Water,* known more familiarly as the *Limited Nuclear Test Ban Treaty.* Atmospheric bomb testing was banned, and subsequent tests were conducted underground. With the collapse of the Soviet Union and the end of the Cold War has come a significant and very welcome movement away from nuclear weapons. This has included a decrease in the testing of nuclear weapons and also in the size of the nuclear arsenals. The *Non-Proliferation Treaty* has been extended for an indefinite period and a *Comprehensive Test Ban Treaty* (including underground tests) has recently been adopted by the United Nations. Tests in the Pacific and elsewhere have ceased. Latin America and Africa are nuclear weapons-free continents, and the United States and the Russian Federation are vigorously downsizing their nuclear arsenals. An unfortunate development that runs counter to this trend is the underground bomb tests conducted by India and Pakistan in 1998.

The average annual dose from atmospheric weapons testing conducted between 1945 and 1963 has decreased from a maximum of about 0.15 mSv per year in 1963 to less than 0.005 mSv per year today. The latter amount represents a negligible contribution (less than one quarter of one percent) to natural radiation.

The notorious Chernobyl nuclear reactor accident in the Ukraine on April 26, 1986, released large amounts of radioactivity into the atmosphere. Air currents carried the cloud in two directions, reaching Canada's east coast on May 6 and the west coast on May 7. The highest air concentrations in Canada occurred in May and by the end of June the levels had returned to normal. Health Canada estimated that the Chernobyl accident led to the typical Canadian receiving a radiation dose of 0.00028 mSv, a negligible amount compared to typical variations in natural background.

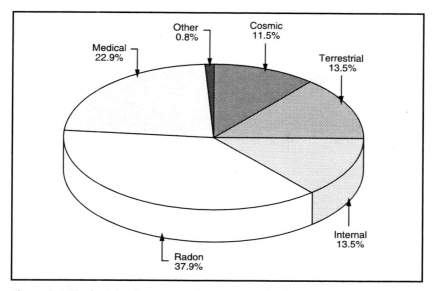

Figure 3-4: Pie chart showing sources of average annual radiation exposure.

SUMMARY

As discussed in this chapter, humans live in a sea of natural radiation. We have seen that the amount of radioactivity any specific individual is exposed to varies with location, altitude, diet, and life style. On average, each Canadian receives approximately 2.6 mSv of radiation per year. This amount of radiation is a useful yardstick and will be referred to as the "annual background dose".

The sources of background radiation are divided into those of cosmic, terrestrial, and internal origin. Radon gas, although of terrestrial origin, is usually discussed separately because it makes such a significant contribution. In addition, the average Canadian receives about 0.6 mSv of radiation dose from dental and medical X-rays and procedures. The contributions to total background radiation are summarized in Figure 3-4, which shows that about 76% (2.0 mSv) is from natural radioactivity and about 23% (0.62 mSv) is from medical (including dental X-rays) and other sources. The category "Other" includes consumer products, fallout from nuclear weapons testing, nuclear power reactors, and uranium mining.

Chapter Four

Biological Effects of Radiation

We are immersed in radiation. Our bodies are radioactive, as are the rocks, plants, and air around us. Humans, and all other living things, have always existed in a sea of radiation; it is a natural and unavoidable part of the world. As a typical adult is 'struck' by about one billion individual radiations each day, it is clear that the cells that compose our bodies can cope with and, indeed, thrive in this radiation field. This is supported by the fact that there is considerable variation in natural radioactivity with some localities having significantly higher radiation than the average. Extensive studies have shown that the people in these areas are just as healthy as those living elsewhere.

Nevertheless, larger exposures to radiation can be damaging to health, as for example, noted in the early 1900s when radium dial painters inadvertently swallowed appreciable amounts of radium during their work. In extreme cases, a very large amount of radiation can cause death, although such a large accidental exposure of radiation has never happened in Canada. Thus, questions in radiation biology do not focus on large doses but instead are directed at revealing what effects small amounts of radiation have on humans. This has important ramifications and, for example, would assist in setting allowable levels of radiation for workers at nuclear facilities. Unfortunately, this is not an easy question to answer and, in spite of considerable efforts, there is some debate on this subject.

This chapter focuses on what are called "low levels" of radiation, that is, radiation doses from 0 to 100 mSv. It should be noted that even though these doses are termed "low level", the upper end of this range greatly exceeds virtually any radiation levels that Canadians would ever be normally exposed to. The natural background radiation delivers on average an annual dose of between 2.0 and 3.0 mSv. These background doses are at the low end of the range referred to as low-level radiation.

Numerous statistical comparisons have been made of groups of people who have received low levels of radiation. No clear evidence has yet been found that such radiation has an adverse effect on human health. Nevertheless, scientists continue to search for low-level radiation effects not only because of the possible implications for health but also because of information they reveal about fundamental aspects of cell biology, for example, **DNA** repair.

There are two different ways of approaching this topic. One is to study radiation effects on the microscopic level to see how ionizing radiation interacts with the cell and different **molecules** within it. The other approach is to look at the problem from a larger perspective in terms of the effects of radiation on populations exposed to different amounts of radiation. Let us look at the former, more detailed approach first.

THE MICROSCOPIC VIEW

Measuring Radiation: Dose Units

To discuss the biological effects of radiation meaningfully, it is helpful to understand the units in which radiation dose is measured. These were mentioned in Chapter 3 and Appendix A and will be described in more detail here.

Radiation deposits energy in the material through which it passes. The energy absorbed by the material is measured using a unit called the **gray** (1 gray equals 1 joule of energy absorbed per kilogram of matter). Thus, the radiation dose to humans could be measured in grays.

If an alpha, a beta, and a gamma each deposit one gray of energy in human tissue, will they have the same biological effect, that is, cause the same amount of damage? The answer is no. When the energy is spread over more cells and molecules, then there is less effect, i.e. less irreparable damage. The alpha particle (helium nucleus), due to its size and charge, travels the least distance so it deposits its energy over the shortest distance. Because this energy is very condensed, spread over a millimetre or less, it causes more damage than the same amount of energy deposited by a gamma, which may be spread over a metre. To allow the biological effects from different radiation to be compared in a consistent manner, a new unit was developed called the sievert (Sv). A sievert is defined as a gray multiplied by a quality factor, Q, which measures the relative biological effectiveness of the different types of radiation. (The quality factor is also referred to as the Radiation Weighting Factor).

The values of Q are:

Q = 1 for gamma rays and beta particles
Q = 3 to 10 for neutrons
Q = 10 to 20 for alpha particles

In other words, the sievert (Sv) is a measure of the energy deposited by radiation in the human body, adjusted for different types of radiation causing different damage. As doses encountered in real life are much smaller than a Sv, the unit millisievert (mSv, one-thousandth of a sievert) is generally used. Typical doses that you, I, and even workers in nuclear power plants are exposed to range from a fraction of one mSv to a few mSv.

There is another complication. In many situations the radiation to a person is not uniform but instead is concentrated in a particular organ, such as the lungs or the hands. Different organs respond differently to the same amount of radiation. Thus, the same dose to the lungs would be quite different from the same dose to the thyroid, because the lungs are 40 times more sensitive to radiation than the thyroid. To allow meaningful comparisons, it is useful to convert an organ-specific dose to an equivalent dose that would be delivered to the whole body to achieve the same effect. In this book we refer only to the whole-body effective dose, even if the dose is to a specific organ.

Radiation can cause damage to living organisms. A schematic illustration of this process is shown in Figure 4-1. As radiation passes through human tissue it knocks electrons out of their electronic shells, leaving a trail of **ion** pairs. That is, the molecule with the removed electron is positively charged, and the molecule to which the removed electron attaches itself becomes negatively charged.

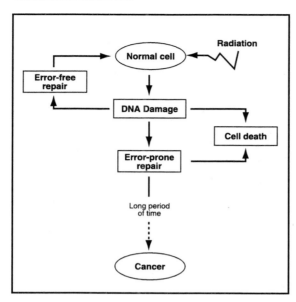

Figure 4-1: Schematic of cancer induction by radiation. Radiation initiates a defect in the DNA of a cell. Usually the damage can be repaired by the cell's own defence mechanisms but sometimes the damage is such that the cell cannot survive or the cell dies as part of a natural mechanism that cleans up defective cells. In rare cases the repair contains errors which transform the cell into a cancer cell. (After R. Mitchell)

These ionized molecules, i.e. with electrical charges, are chemically very reactive and want to form new molecules. The resulting chemical changes can upset the natural chemical reactions that go on in living cells and can cause biological effects.

The human body is composed of billions of cells formed into organs and related systems to support life. The different types of nuclear radiation have different effects on the body and its organs and tissues. Gamma rays can penetrate deep into the body and cause damage even from a distance. Beta particles can only penetrate a short way into tissues, nevertheless, they can cause skin burns and damage if ingested. Because alpha particles cannot penetrate the skin, they are a health concern only if they are carried inside the body by breathing or eating.

The effect of radiation on the human body depends on the duration of the exposure, how much energy is deposited in the body (i.e., the strength and type of radiation), and the type and number of cells that are affected. The body is by no means defenceless against radiation and cells have repair mechanisms.

Through studies of 96,000 survivors of the 1945 atomic bombings of Japan, scientists have learned a great deal about the biological damage caused by large radiation doses given over a short time (known as acute exposure). In large doses, ionizing radiation can cause death or damage to cells and cell components and, when sufficiently large, they can cause the organism to die. For example, a radiation dose over 1000 mSv would usually cause radiation sickness and a dose of 4000 mSv can cause death. Such acute exposures are extremely rare, for example, such as suffered in the Japanese atomic bombs and by some firefighters and reactor operators at the Chernobyl accident. No acute exposures of this magnitude have ever occurred in Canada by accident (although controlled large doses are used in cancer therapy).

Chronic exposure consists of low levels of radiation over long periods of time. For such chronic doses, there are no immediately observable effects. Any effects, if they do occur, appear many years later and, thus, are difficult to link with the original cause. Because of this uncertainty, considerable controversy has arisen over low doses of radiation.

Although radiation can cause minor damage to other molecules, it can have a significant impact only on a special molecule called DNA, which contains information that controls how that cell behaves (i.e., genetic material). Ionizing radiation can cause damage to the DNA molecule that may affect its future behaviour. It is unrepaired or incorrectly repaired cell damage that is critical, rather than cell death. A dead cell is replaced, but a damaged cell may replicate itself. For example, if the cell multiplies out of control, it may become a **cancer** tumour. Organs with rapidly dividing cells (bone marrow, gonads, intestines, growing children, babies in a womb) are most susceptible to radiation damage. For this reason, pregnant women are not allowed to work within the radiation-containing areas of nuclear facilities.

If the damaged cell is a sperm or egg cell that subsequently develops into a fetus, hereditary effects may occur. Although this has been observed in animals, radiation-caused **genetic defects** have never been observed in humans, despite extensive studies. For example, no significant increase in genetic disorders has been detected in the children of the Japanese bomb survivors at Hiroshima and Nagasaki.

Somatic effects are those effects that appear in the exposed person. Genetic effects are those that appear in children after the parent has been exposed.

RISK

Nothing is perfectly safe; all activities in life carry a risk, whether it is driving a car, crossing a street, smoking a cigarette, climbing a mountain, or enjoying a swim. When regulatory agencies such as Environment Canada, Health Canada, and the US Environmental Protection Agency set limits on, for example, the maximum amount of certain heavy metals allowed in drinking water, they do so by calculating the risk that ingestion of that chemical would cause. The maximum concentration is set so that the resultant risk of death or serious health damage is very small (it is impossible to make it zero). The risk number is determined by comparison to other risks that humans accept, such as the risk of an accident on an airplane trip, and so on. A standard that has evolved in recent years is that such risks should be kept smaller than about one chance in 100,000 to one chance in 1,000,000 over a lifetime. To get a perspective on this, imagine 100,000 thimbles on a (large) table; only one thimble has a pea under it. Given only one choice, the chance of your selecting the right one is one in 100,000.

Some typical risks in everyday life are as follows. The risk of death from a driving accident is one in 5,000 per year or one in 100 over a 50-year period. The risk of death per year from an accident at home is one in 11,000. The risk of death per year from an accident at work is one in 24,000. The probability of hitting the jackpot in Lotto 649 is about 1 in 14 million.

The main adverse effect of radiation on humans is cancer, which does not generally become evident until many years or decades after the exposure. Extensive studies with survivors of the atomic bomb detonations in Japan show that no other diseases are caused by radiation. In fact, at low doses, no cancer has ever been positively correlated as being caused by radiation. Many people do get cancer, however, and it is thought that some small fraction of those may come from radiation.

Laboratory Experiments
One might think that exposing bacteria, laboratory animals, and human cell colonies to low-level radiation in the laboratory might provide answers. Some experimental results have been as follows.

• Bacterial colonies exposed to three times the natural level of radiation thrive much better than those exposed to less than the natural background level.

• Various strains of mice exposed to very low doses of gamma radiation live longer and have fewer cancers than non-irradiated mice. The survival time of 50% of the exposed mice was 673 days, compared with 549 days for the control group. Furthermore, exposed mice had fewer cancers than their non-exposed counterparts.

• Tests with cell colonies show that at an average of one radiation hit per cell (approximately the number of radiation hits per year in human cells at natural radiation levels) the risk of transformation of a cell to a cancer cell was reduced three to four times below the rate of transformation when there was no radiation. The data also show that even higher doses, up to 100 times greater, but delivered at a slow rate, produced the same reduction in cancer transformation risk.

These results are unexpected and difficult to understand.

Search for a Risk-Dose Relationship
In view of these laboratory findings, there is some controversy among scientists about the effects of low exposures of radiation. Virtually all debate about health effects of radiation centres about what is called the "linear no-threshold" dose theory. It is a bit complicated, but provides a fascinating insight into statistics and science.

Ideally, medical scientists would like to know the exact relationship between dose and risk, that is, if a person receives a certain radiation dose (in mSv), what is the risk of getting cancer (usually expressed in probability over a lifetime).

The dose range of interest is 0 to about 100 mSv, which includes the doses expected under almost all conditions, except the most severe reactor accidents. Such a relationship would be valuable in determining regulatory standards and generally assisting in radiation protection. That is, standards would be set for nuclear facilities and practices so that the doses to their workers and the public would not exceed an acceptably small risk, usually about 10^{-5} to 10^{-6}, that is one chance in 100,000 to one chance in 1 million. In spite of decades of intensive research, it has proven very difficult to determine such a dose-risk relationship.

Figure 4-2 shows Cancer Risk plotted against Radiation Dose. As expected, the higher the dose, the greater the risk of getting a cancer. The data to plot this graph have all been obtained from high doses, primarily from the nuclear bomb detonations at Nagasaki and

Hiroshima, the reactor accident at Chernobyl, early X-ray workers, early radium dial painters, bomb testing at Bikini Atoll, and also from animal tests.

The curve for the low-dose section in which we are interested (0 to about 100 mSv) is not known. It is exceedingly difficult to obtain reliable dose-risk data for this part of the graph because, as discussed earlier, there are no immediate biological effects of low radiation dose, and any effects, if they occur, do so many years after the dose has been received. Cancer is primarily a disease of old age, and may occur 10 to 40 years after the dose has been received. Furthermore, if a cancer does appear, it is impossible to separate radiation-caused cancers from those generated by other causes. More than 1,500 different agents other than radiation are known to cause cancer, including arsenic, tobacco smoke, asbestos, ultraviolet light, paraffin oil, fungal toxins in food, viruses, and even heat.

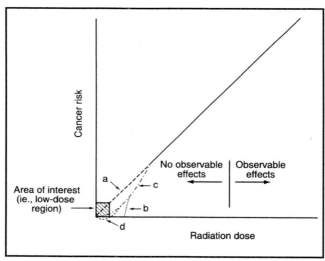

Figure 4-2: Curves illustrating various radiation-dose versus cancer-risk models used for low radiation exposures: (a) linear, (b) threshold, (c) S-shaped, and (d) hormesis.

Because of the difficulty in accurately correlating dose and small effects that may or may not occur years later, it is easy to misinterpret data. Suppose, for example, that 20,000 cancer deaths are predicted in a population of 100,000 people. What is the significance if 20,300 deaths were actually to occur? Does this mean that radiation, or some other cause, led to 300 excess deaths? Or is this simply part of the natural statistical variation? This uncertainty would have been removed if the predicted cancer deaths had been quoted together with the expected variation, as 20,000 plus or minus 1,000. Because of this type of statistical variation, it is possible to "prove" almost

Table 4 -1	CANCER DEATHS AND THEIR CAUSES	
	Cause	Cancer Deaths (%)
	Diet	35
	Tobacco	30
	Infection	10
	Sexual lifestyle	7
	Occupation	4
	Alcohol	3
	Natural environment	3
	Pollution	2
	Medical care	1
	Food additives	1
	Industrial products	1
	Unknown	3

anything. Data of this nature should be carefully reviewed by experts in statistical analysis and epidemiology.

In western countries about 20% of all deaths are due to cancer, with the average cancer death occurring at about 65 years age. Table 4-1 shows the various causes of cancer deaths. It is seen that 3% of cancer deaths are attributed to the natural environment, which includes natural radiation. Diet (35%) and smoking (30%) are the main causes of cancer deaths in the US and in Canada.

EPIDEMIOLOGY

Epidemiology is the science of epidemics, or the study of diseases that attack many people. It is particularly useful when there is no readily observable relationship between the cause and symptom, so that large numbers of people must be studied, generally over a long period of time, using computer databases and statistical analyses.

For the preceding reasons, the shape of the dose-risk curve for doses below 100 mSv is not known. Nevertheless, various hypotheses have been proposed. Four different forms of the dose- risk relationship are discussed below and illustrated in Figure 4-2.

a) The most widely used theory assumes a straight line all the way from high doses to zero dose, as shown by curve "a" in Figure 4-2. This is the so-called Linear No-Threshold (LNT) theory. The line is based entirely on data obtained at doses that are far outside the region we are interested in. The slope of the line yields the risk per dose, which is 0.004% cancer risk per mSv of dose. The linear theory has one very important implication: all radiation is harmful, no matter how small the amount.

It is important to stress that the majority of scientists and doctors who have studied the subject believe that low levels of radiation have an extremely low risk, if any, of a harmful effect. In fact, virtually all expert scientific groups recognize that the linear no-threshold theory overestimates risk. For example, both the United Nations Scientific Committee on the Effects of Atomic Radiation (1988) and the US National Committee on Radiological Protection and Measurements (1980) estimate that, when radiation is delivered at low dose rates, i.e., spread over a long period of time, risks at low radiation levels are from two to 10 times less than would be predicted from the linear theory. The US Committee on Biological Effects of Ionizing Radiation (1990) also acknowledges that the linear theory overestimates risk, but does not say by how much.

Although the linear theory does not match what is actually observed, it continues to be used by regulatory agencies to set radiation dose regulations because it is conservative. That is, regulations based on the linear theory will have an extra factor of safety. The linear theory also has the advantage of being simple and easy to administer. For example, even though many scientists believe that the actual relationship includes a threshold (curve b), determining the specific value for this threshold is very difficult.

b) The most likely possibility is that there is some threshold value of dose below which there is no risk, as shown by curve "b" in Figure 4-2. Only doses that exceed the threshold value cause cancer risk. This curve matches evidence that our bodies can repair and cope with small amounts of damage. It is also consistent with how our bodies deal with many other substances. For example, molybdenum and iron are vital ingredients for health in trace amounts, yet are poisons if taken in large amounts. Another example is the common headache tablet, which if taken in large quantities can be lethal, but if taken by ones and twos eases pain and headaches. Furthermore, an exhaustive study of workers who painted radium dials was conducted over 40 years and showed that there was a dose threshold.

c) Another possibility is that the risk/dose curve lies somewhere between the linear and threshold curves. This is a compromise between the linear and threshold theories. The curve could take many forms of which the S-shaped curve (curve "c" in Figure 4-2) is used relatively often. This curve indicates that the risk at low doses is quite small but is not zero.

d) A fourth alternative is that low radiation doses can actually have a beneficial effect. This effect is called **hormesis** and is similar to immunization and vaccination, where small doses provide protection against larger doses. A substantial body of evidence is accumulating that cells exposed to low-level radiation suffer less damage later when exposed to larger levels of radiation. For example, it has been observed that Japanese bomb victims who received only small radiation doses, are getting fewer cancers and living longer than the control group who received no bomb dose. This effect has been traced to stimulated production of repair enzymes by low-level radiation. Hormesis is illustrated by curve "d" in Figure 4-2.

Let us see what happens when the linear theory is applied to head-ache tablets (or copper or many other chemical compounds). Suppose that 20 aspirins, if ingested at one time, would cause death. If this data were plotted on a risk-dose graph, a straight line could be drawn from the data point to the origin. The resulting risk-per-tablet ratio is 0.05, that is, a linear no-threshold theory would predict 0.05 chance of death per tablet. Suppose that in North America 300 million people each take one tablet per year, on average. The linear theory then predicts that 15 million people would die each year from taking the head-ache tablets (300 million people x 1 tablet per person x 0.05 risk of death per tablet). We all know such deaths do not occur and, thus, a realistic model would have a threshold. In fact, one aspirin a day is thought (and has been advertised) to reduce the risk of heart attack, thus, having a beneficial effect — an example of hormesis.

The linear no-threshold dose-risk model is useful for regulatory agencies because of its simplicity, and because its application in deriving regulatory limits provides a large safety margin. Because this model overestimates risks, its application to other areas should be avoided.

POPULATIONS: THE MACROSCOPIC VIEW

Let us now step back and look at the forest, rather than the trees, that is, let us inspect the biological effects caused by low levels of radiation from a broader perspective.

In Chapter 3 we saw that every part of the earth is radioactive and that humans have evolved with radiation as an ubiquitous part of the environment. Everything contains radioactivity: the rocks and soils around us, the bricks and concrete from which our homes are built, the food we eat, the air we breathe, and even our own bodies. It is impossible to find a place on earth or to construct one — no matter how ingeniously — that would be totally free of radiation.

As Charles Darwin stated in his theory of evolution, living things evolve through a process of natural selection and develop defence mechanisms against detrimental stimuli in the environment. Thus, it is unimaginable that living organisms would not have evolved and adapted to the levels of radiation encountered in nature, especially since natural radiation was even higher in earlier millennia.

The process of natural adaptability is well illustrated by bacterial diseases and their resistance to antibiotics. Today, we face a potentially serious problem due to the genetic versatility of bacteria. Every major disease-causing bacterium now has strains that resist at least one of the over one-hundred antibiotics used to treat them. Scientists estimate that bacteria typically become resistant in 5 to 10 years to new antibiotics that today are taking at least one decade to develop. Thus, we are only postponing the inevitable: bacteria will always have the ability to adapt rapidly (on a human scale) to any antibiotics we develop. Mosquitoes and insects are also excellent examples and adjust within a few decades to become immune to DDT and other lethal pesticides that have been engineered specifically to eradicate them.

If bacteria and mosquitoes can adapt to various chemicals, so can humans. Thus, human cells have developed defence mechanisms and adapted to cope with radiation, an environment to which they have been exposed for countless generations over millions of years. The continued existence of harmful bacteria and mosquitoes are living proof that natural levels of radiation are not harmful to humans.

Locations with High Radiation

Radiation is not constant, but varies from place to place in the world. As noted in Chapter 3, people living in high-altitude cities such as Banff, Mexico City, and Denver receive more cosmic radiation than those living near sea level. High levels of radioactivity also occur in areas with granite outcrops such as Cornwall, England, and in the Precambrian shield areas of central and northern Canada. Studies have shown that people living in areas of higher natural radiation are just as healthy as those living elsewhere.

For example, millions of people have lived all or large parts of their lives in Colorado, where they receive more natural radiation (both cosmic and terrestrial) than people living elsewhere in the USA. If low level radiation was grossly damaging to health then the citizens of Colorado should have greater mortality, particularly from cancer, than those in other parts of the USA. Studies have shown that exactly the opposite is true: the cancer rate in Colorado is 35% below the national average.

There are many similar examples. A high natural-radiation region in China has radon levels three times higher than average. Han peasants have lived there for 600 generations, yet they show no difference in lung cancer mortality, frequency of hereditary diseases, or any other cancer mortalities compared to the people living in other areas. Kerala, India is underlain by radioactive sand that results in local inhabitants receiving four times the natural radiation dose compared to the rest of India. A survey of 70,000 people over a 35 year period showed no increase in genetic or other adverse health effects.

These observations suggest that the linear no-threshold model should be replaced by a threshold model (curve b in Figure 4-2).

A strong case for the hormesis model was presented in a comprehensive study of lung cancer mortality and average radon concentration in homes. This study included 1,601 US counties, representing 90% of the US population, and concluded that, with or without corrections for smoking prevalence, there is a strong tendency for lung cancer rates to decrease with radon exposure. In spite of extensive efforts, no explanation for this result could be found other than the failure of the linear theory at low radiation levels.

The average natural background radiation (external sources alone) in all US states is 1.3 mSv, and in the seven states with highest natural radiation, the average is 2.1 mSv. Yet the cancer rates for all malignancies were lower in the seven states with high radiation background (126 versus 150 per 100,000 per year) and lung cancer rate was 14.5 versus 20.4 (per 100,000 per year).

Table 4-2 lists a few of the locations in the world where elevated levels of natural radiation occur and shows the doses that the local

Table 4-2:	LOCATIONS WITH HIGH NATURAL RADIOACTIVITY	
	Location	Annual Dose (mSv/year)
	Finland (average)	8
	US Rocky Mountain States	6 to 12
	Haute Provence, SW France	88
	Kerala Coast, India	5 to 70

inhabitants would receive. Note that these dose levels significantly exceed the average natural dose received by Canadians of 2.6 mSv per year. If radiation doses of these levels were encountered near nuclear facilities or around nuclear accidents, the media uproar would be sensational and towns would undoubtedly be evacuated. Yet the evidence gathered to date shows that the inhabitants of these areas live normal lives with no signs of adverse effects.

In common with the laboratory results, the data on populations seem to indicate that any effect that low level radiation might have on populations is either too small to measure or goes contrary to the expectation that low radiation levels are harmful.

Occupational Doses

Because radiation is readily detected, it is easy to monitor the radiation doses received by workers at nuclear power plants and research institutes to ensure there are safe working conditions. People at such facilities wear personal measuring devices called dosimeters. Direct-reading and thermoluminescent dosimeters are the most common.

The direct-reading dosimeter allows the wearer to periodically measure the amount of gamma radiation present at that time. A quartz fibre within the dosimeter measures the radiation and moves along a calibrated scale to provide a reading. More modern electronic dosimeters give instantaneous readings of dose and dose rate.

A thermoluminescent badge (Figure 4-3) contains a lithium fluoride crystal which records the amount of gamma radiation to which the worker has been exposed over a period of time. When heated, the crystal gives off an amount of light (luminescence) that is proportional to the dose received. This allows automated methods of reading and recording the data from the badges. Compare this to the difficulty of monitoring exposure to non-radioactive contaminants such as might occur, for example, in pesticide factories.

A new personal alpha dosimeter has been recently introduced for use by uranium miners. Direct-reading and thermoluminescent dosimeters measure gamma radiation, whereas in a uranium mine the principal risk comes from radon gas, which results in lung exposure to alpha particles. The new dosimeter contains a small battery-operated pump that draws air through a special filter which is sent monthly to a laboratory for counting. More than 500 Canadian uranium miners are monitored by this device.

The Canadian Nuclear Safety Commission sets the exposure limits for what are called Nuclear Energy Workers, which includes employees at nuclear reactors. Whole body radiation exposure is not to exceed an annual limit of 20 mSv, averaged over five years.

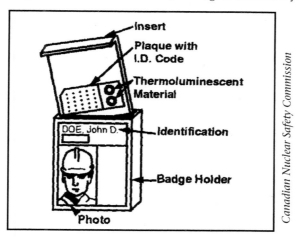

Figure 4-3: Thermoluminescent radiation badge carried by workers at nuclear installations. The thermoluminescent material is removed periodically and "read" to measure radiation exposure. Usually the badges are also used for identification and carry the person's photograph.

Nuclear workers are at the greatest risk from radiation since they often handle or are in close proximity to radioactive materials. Employees at nuclear reactors are carefully monitored. They wear thermoluminescent badges when at work and also have regular radiation measurements of urine as well as whole-body counting. In addition, radiation monitors are located throughout the reactor building to detect radioactive contamination. From 1955 to 1987, during more than 133 million worker-hours of research and commercial operation of Ontario Power Generation nuclear facilities, not a single employee was injured due to radioactivity and there were no medically-significant radiation exposures.

A mortality study from 1971 to 1989 of all male workers and pensioners of Ontario Power Generation showed that 34% fewer nuclear station workers died during this period than would be expected statistically, based on male mortality rates in Ontario. There were also fewer cancer deaths than expected. These results are due, at least in part, to the close medical attention and study the workers receive.

A 29-year study tracked US Army radiology technicians who received 500 to 1000 mSv doses over a three-year training period in World War II by taking large numbers of X-rays of each other (these are much larger doses than any member of the public would ever receive). There was no difference in cancer mortality among these technicians and a control group of medical technicians.

THE CHERNOBYL ACCIDENT

As is discussed in Chapter 8, there has only been one reactor accident with a large radiation release to the public, namely the one in 1986 at Chernobyl in the Ukraine. At the time of the accident 28 firefighters and reactor operators lost their lives from high radiation doses, and three others died of injuries. There has been considerable debate, however, about the effect of radiation exposures on the public over the longer term.

Huge amounts of radioactive materials were put into the atmosphere — up to 12 trillion becquerels — much of it released in the first ten days of the accident when the reactor was burning. This amount is about 400 times more than the radiation released by the Hiroshima bomb, and between a 100 to 1000 times less than the total radioactive materials released by the nuclear weapons tests of the 1950s and 1960s. Although hundreds of radioisotopes were released, the main radiation exposure to humans and the environment arose from isotopes of cesium, iodine, and strontium. Radioactive materials were deposited in patches depending on local weather conditions, with some locations far from the reactor having higher levels of radiation than others near it. The radioactivity from the accident was deposited mostly in the Ukraine and its neighbour Belarus, with the largest concentrations near the reactor. Much lower levels of radiation were experienced in other countries in the northern hemisphere. The maximum exposure to an individual outside the then Soviet Union was 0.8 mSv, about one-third of one year's natural background exposure.

A special group of policemen, firemen, soldiers, and volunteers

IN CASE OF ACCIDENT
GO TO THE GAS METER

Potassium iodide or iodate tablets can protect people, particularly children, against thyroid cancer caused by a radiation accident. These pills flood the thyroid with non-radioactive iodine so the gland has no capacity to absorb any additional radioactive iodine. Iodine tablets were provided to some 5.3 million people during the Chernobyl accident of whom 1.6 million were children. For this treatment to be most effective the pills must be taken before the radioactive iodine is absorbed.

Ideally, the tablets should be stored in every house, school and office near reactor installations where they would be available for immediate use in the event of an accident. Like most other types of pills, they deteriorate over time losing their effectiveness and should be replaced periodically. In North America iodine tablets are generally stored in large quantities at central locations, but they are not replaced regularly.

Some years ago in the UK iodine tablets were kept up to date by gas company employees who entered homes to read the gas meters. They would leave the pills on top of the gas meters where families knew to go in case of a nuclear accident.

were called upon to put out the reactor fire and then to clean up the accident. It is estimated that there were about 200,000 of these so-called "liquidators". This number grew to between 600,000 and 800,000 in the years after the accident but many of this additional number received only very low levels of radiation.

The average radiation exposure of the initial 200,000 liquidators was about 100 mSv. Approximately 20,000 of them were exposed to 500 mSv, and a few hundred received thousands of millisieverts. As a consequence of the accident, 237 people were admitted to hospital, 134 of them with acute radiation syndrome. Of this group, 28 died within three months of the accident and at least 14 others within about ten years, although it is not clear whether radiation was a factor in the deaths of this latter group.

All the civilian population living within a 30 kilometre radius of the plant were evacuated, about 116,000 people in all. Pripyat, a city of 45,000 people, was totally abandoned . Of those evacuated, it is estimated that about 10% received exposures greater than 50 mSv and less than 5% received exposures exceeding 100 mSv. Later, an additional 210,000 people were evacuated from high-radiation areas in the Ukraine, Belarus, and Russia out of a population of just over a million.

In the years following the accident, the major radiation-induced illness has been an increase in thyroid cancers, particularly in children. Cancer was caused by inhalation of iodine-131 at the time of the accident or by drinking milk from cows eating contaminated grass. Iodine-131 has a half life of only eight days and was essentially gone in three months. Children born after that period have the normal incidence of thyroid cancer. It is estimated that at least 1,800 additional cases of childhood thyroid cancer occurred in the ten years following the accident. Fortunately, this type of cancer is relatively easy to detect and can be cured by removal of the thyroid and, thus, most of these young victims have survived, although the survivors must take thyroid hormones for the rest of their lives. It is estimated that there will also be up to a few thousand adult cases.

If the linear hypothesis is applied to the radiation releases from Chernobyl, we would expect about 100,000 deaths in a population of a few million. Medical teams sent in by the United Nations World Health Organization and the International Atomic Energy Agency have been looking for fifteen years for excess mortality in the population exposed to Chernobyl radiation. The people exposed have been very carefully monitored, their health has been scrutinized, and their lives placed under a microscope. So far no increases have been detected in the incidence of leukemia or other cancers. As the latent period for some of these cancers can be twenty years or more after the radiation exposure, these populations will need careful monitoring for decades to come. Nevertheless, to this point in time there is no sign of the large number of deaths predicted by the linear hypothesis. Similarly, no increases in genetic defects have been observed in children born in the years following the accident.

Even in the case of an accident as severe as Chernobyl, humans seem to be remarkably resilient to radiation. Humans have always been exposed to low levels of radiation and our bodies have had millions of years to evolve and adapt to this natural phenomenon. This may explain why so far, in spite of extensive monitoring, there has been no convincing evidence for harmful biological effects caused by long term exposure to low levels of radiation.

Chapter Five

Electricity – How Did We Live Without It?

An old man, who had just reached the venerable age of 100, was asked to describe the most important change he had witnessed during his life. Without hesitation he said, "The introduction of electricity. I don't know how we lived in the days without it."

Providing a reliable electricity supply is one of the biggest challenges facing modern society. Our worst nightmare is to have power failures or to suffer brownouts. We only need to look at the human suffering and economic devastation that was caused by the ice storm of 1998 in eastern Canada. Electrical transmission was severely disrupted when poles and wires snapped under the weight of the ice, leaving large segments of the population without electricity in the middle of winter, some for more than three weeks. The rolling blackouts in California in early 2001 due to lack of electrical-generating capacity also caused considerable stress.

Our society thrives on electricity; it is a fundamental part of our homes, businesses, and factories. When we turn the switch, we expect the light to come on, the kettle to boil, and our shavers, vacuum cleaners, irons, hair dryers, TV sets, computers, washing machines, power tools, and innumerable other items to work. This is all possible because electricity transports energy into our homes and offices. There is hardly a facet of modern society that is not in some way connected to the electrical grid; we are dependent upon a reliable source of electricity.

ENERGY AND ELECTRICITY IN SOCIETY

Globally, electrical energy use is growing at an enormous rate. In 1990, for example, nuclear reactors produced as much electricity as was generated in the entire world from all sources only 25 years earlier. This demand is fuelled by two fundamental and irreversible dynamics, namely, the growth in population and the desire for a better standard of living.

In the early days, human population was small and the world must have seemed an infinitely vast place. But as the population increased to a few hundred million and from there to a billion and then, ever more rapidly, to the present six billion, the globe has become crowded. The enormous population increase is further complicated by the trend to urbanization; we are living in ever more dense and concentrated conditions. In 1984 only 34 cities had populations exceeding five million; by 2025, the United Nations estimates there will be 93 cities of this size or larger. Enormous quantities of energy are needed to provide necessities such as sanitation, heating, cooling, lighting, and transportation for such large, high-density populations.

In the earliest days, humans used wood as a fuel. Then the burning of coal ushered in the industrial revolution, at the same time as the rapidly expanding population denuded the forests of Great Britain and much of Europe. In the past century we have begun to burn oil and natural gas to produce power. Humans have also harnessed renewable energy sources, in particular, falling water but also the sun, wind, and tides. Nuclear energy was introduced as a power source following the Second World War. The progression from wood to **fossil fuels** (coal, oil, natural gas) to hydro electricity to nuclear represents a steady evolution from low technology to more complex technologies. As energy resources become scarcer, it requires more ingenuity to extract them.

Electrical power is a fundamental part of the aspirations of third-world countries. China, for example, has the ambition to ensure that every household has a refrigerator. Even if they were the newest most energy-efficient models, these refrigerators would require an electrical supply of 20,000 MW, or about 25 nuclear reactors of the size of those at Darlington. And this does not take into account stoves, televisions, other appliances, or the power needed by the factories to manufacture these goods. Projections indicate that by 2040 the world will need three to four times the electricity capacity that we have today — an incredible increase. It is in the face of this enormous growth in demand for energy that we must view the need for nuclear power as an energy source.

The demand for electricity is growing faster than for other power sources because of its convenience as well as its environmental friendliness. The use of trolley cars and subways instead of diesel buses is an example. Furthermore, battery-driven cars and electrical trains are the transportation methods of the future. Thus, electrical demand will continue to grow dramatically in both third-world and industrialized countries.

Canada's population is steadily increasing, as is the use of electricity and other energy sources. Canada has a greater need for energy than other countries because of the cold northerly climate, our high standard of living, our geography, which spans long distances, and basic industries that rely on processes that use a great deal of energy such as aluminum smelting and pulp-and-paper manufacturing.

This demand can only be satisfied by energy megaprojects such as the enormous tar sands projects in northern Alberta that produce oil; pipelines thousands of kilometres long that carry oil and natural gas from western Canada to consumers in eastern Canada and the USA; large and complex drilling platforms in the Hibernia oil fields off the coast of Newfoundland; and vast water reservoirs that have inundated millions of acres in northern Quebec to provide hydroelectricity. Compared to these methods of energy production, nuclear reactors are relatively small and compact.

Canada is fortunate to have a variety of energy sources available in relatively plentiful supply and at affordable prices. Electricity accounts for about 20% of all energy use (see Table 5-1) with the remainder used to fuel cars, trucks, and aeroplanes, to run factories, and to heat our homes and offices in the long Canadian winter.

Table 5-1	PRIMARY ENERGY IN CANADA IN 2000		
	By Fuel Type	By Sector	
	Oil 40 %	Industrial	43 %
	Natural Gas 27 %	Transportation	25 %
	Electricity 20 %	Residential	19 %
	Renewables 7 %	Commercial	13 %
	Other 6 %		

Table 5-2	ELECTRICITY CAPACITY BY "FUEL" TYPE IN CANADA	
Fuel type	1995	2010
Hydro	64,500 MW (59%)	67,000 MW (59%)
Coal	17,000 MW (16%)	16,000 MW (14%)
Oil	7,500 MW (7%)	6,000 MW (5%)
Nuclear	14,000 MW (13%)	13,000 MW (11%)
Natural Gas	4,000 MW (4%)	8,000 MW (7%)
Other	2,000 MW (2%)	4,000 MW (4%)
Total	109,000 MW	114,000 MW

* GW = Gigawatt = a billion Watts = 10^9 Watts

Table 5-2 shows the different methods used to produce electrical energy in Canada in 1995, and their projected use in 2010. Nuclear accounted for about 13% of electrical power in Canada in 1995. This compares to the world average of about 19%. (For the percentage contributions of nuclear to total electrical energy for other countries see Chapter 7, Table 7-3.)

ENERGY CHOICES

The choice of which energy sources to use for generating electricity depends on many factors.

Baseload Versus Peaking Capacity

Electricity, unlike other manufactured commodities, cannot be stored on a large scale; it must be made as it is needed. Thus, what counts is not inventory, but the capacity, or capability, to generate the product. Furthermore, electricity demand is not constant but varies throughout the day, with maximum use in the morning and evening and minimum use during the night; it also varies seasonally. To match this cyclical demand, generating capacity is divided into two categories: base load, which operates all the time, and peaking power, which is turned on and off to meet the cyclical demand.

Because of the complexity of nuclear plants, they are not well suited for turning on and off or for changing power levels frequently. But because of their low fuel costs, they are ideal for providing base-load capacity. Oil-fired plants and natural-gas turbines have expensive fuel but are the cheapest to build and are relatively simple to operate; thus, they are well suited for providing peaking power. Hydro (water) power is also well suited for peaking power.

Resource Availability

Another consideration is the availability of resources. Ontario, for example, has little indigenous coal, so it must be imported from Pennsylvania, making nuclear cost-competitive with coal. Figure 5-1 shows Ontario Power Generation's Lakeview Thermal Generating Station which is fuelled by coal. Alberta has ample low-sulphur coal so nuclear is not cost competitive there. The Maritimes have abundant coal, but with a high sulphur content so that, once pollution-control devices are added, electricity generated by coal and by nuclear are about the same cost.

Only British Columbia, Quebec, Newfoundland (Labrador), and Manitoba have abundant undeveloped hydro resources. In all other provinces, including Ontario, virtually the full potential of hydro-

Ontario Power Generation

Figure 5-1: The Lakeview Thermal Generating Station operated by Ontario Power Generation is located on Lake Ontario at the west end of the Greater Toronto Area. At full capacity of 1,140 MW, the four coal-burning units use more than 8,000 tonnes of coal per day. Note the coal handling facility at the bottom right.

electric power has already been developed. On a world-wide basis, about 60% of the hydro power potential is already exploited, and in industrialized nations the exploitation is close to 100%.

Renewable Energy

The use of wind power has been growing in recent years as the technology of windmills advances. In 1994, twenty thousand turbines worldwide provided 3,000 MW of energy. Of these, 85 per cent were located in California where they produce 1% of electrical power. Denmark is also a big user, with about 3% of their electricity produced by wind. But this seemingly benign energy source is not without its complications. Aside from requiring vast areas of land, windmills kill and maim birds of prey such as hawks, falcons, and eagles, who like to soar on air currents near the same ridges that are best suited for windmills.

Large urban centres with their densely-packed apartments, homes, factories, offices, theatres, shopping malls, and subways consume enormous quantities of energy. Large power plants are needed to supply this demand. Thus, renewable energy sources such as solar and wind power, although excellent sources of clean energy, particularly in rural settings, will not be able to make more than a minor contribution in urban settings. For example, the Tiverton wind turbine (pictured in Figure 5-2) has a capacity of 0.6 MW, but since wind sufficient to generate electricity only blows about 23% of the time, its effective average capacity is about 0.14 MW. This forms a minuscule fraction of the typical electrical energy demand experienced by Ontario Power Generation of approximately 20,000 MW. About 20,000 wind turbines the size of the one at Tiverton covering thousands of acres of land would be needed to generate the same electrical capacity as the Darlington nuclear station. Furthermore, because wind speed is variable and may stop altogether during calms, a backup electricity generating system or an efficient energy storage system is also needed.

Wind turbines have been constructed in Alberta, and two wind turbines are planned for Toronto's lakeshore. They will be about 100 metres high and will produce about 2.5 times the energy output of Tiverton.

Solar energy can be used to generate electricity using photovoltaic cells. Although this is commonly done in small-scale applications such as hand-held calculators, it has not yet been used in Canada on

Ontario Power Generation

Figure 5-2: The wind turbine at Tiverton, Ontario. Built by Tacke of Germany in 1996, the unit has blades 21 metres long on a tower 50 metres high. With an energy output rating of 0.6 MW, it generates enough electricity for about 100 homes.

a scale where the solar electricity is fed into a power grid. Solar energy is also well suited for domestic hot water and heating homes, offices, and swimming pools. Although the technology is available, solar water heating is not yet widely used due to its relatively high cost.

Solar energy has similar limitations as wind power; the energy coming from the sun is diffuse and intermittent. To supply China's 1.2 billion people with solar power using today's technology, for example, would require solar panels with about the same area as Saskatchewan. Even with significant improvements in solar-panel technology, the area of the collection system would be vast. Although solar energy will make a contribution, China has turned to other sources for its primary energy supply. One of the main sources has been coal-fired generation, which has led to serious air pollution problems. China has also chosen to build nuclear reactors.

Renewables should be used to the maximum extent possible, but due to their diffuse nature they simply cannot contribute the large blocks of energy needed by our large and growing population. The current plan for Ontario Power Generation is to have 2% of Ontario's power from renewable energy sources by 2005 — an ambitious target.

Hydrogen and Fuel Cells

Even if we had abundant electricity, there would still be a very large requirement for 'transportation' fuels such as gasoline and diesel fuel to power cars, trucks, aircraft, and other vehicles. As these non-renewable fuels are being depleted, attention is being increasingly focussed on hydrogen and fuel cells to replace them and as the basis for sustainable energy systems.

By passing an electric current through water it is possible to separate the water molecule into hydrogen and oxygen gas in a process called **electrolysis**:

$$\text{water + electrical energy} \rightarrow \text{hydrogen + oxygen}$$

Hydrogen gas is portable and can be used as a 'transportation' fuel. This has been done for experimental cars, buses and even lawn mowers. When hydrogen burns it produces only heat and water and, thus, is non-polluting.

Electrical energy is required to drive the electrolysis process, and this can be supplied by conventional power plants, including nuclear reactors. If hydrogen were to be made from solar electricity, for example, and then burned as fuel in vehicles, it would provide an almost pollution-free situation. At least two small pilot projects of this type have already been tried but were found to be very expensive. Hydrogen is difficult to store and special safety precautions are required in its handling. If hydrogen is to become the fuel of the future, a whole new infrastructure of hydrogen transmission and filling stations will be needed.

The above equation can also go in the opposite direction, that is, hydrogen and oxygen can be recombined to produce electricity. This process happens in a fuel cell, which can be thought of as a special type of battery. The fuel-cell concept was originally proposed in 1839, but the first application was as a substitute for conventional chemical batteries in space. The only waste from the fuel cell is water, so it is environmentally clean. Hydrogen batteries are portable and can be used to power vehicles such as cars, trucks and buses.

Thus, another way of storing energy would be to store hydrogen. For example, excess electricity generated by a solar panel or nuclear reactor could be used for electrolysis. The resulting hydrogen could then be stored and, at a later time, used in a fuel cell to generate electricity.

Hydrogen fuel cell research and development is currently a very active area. The main challenges are to bring down the cost of fuel cells and to increase the electrical power that can be generated by a fuel cell of a given volume (power to volume ratio). The leading fuel cell company in the world is Ballard Power Systems of Burnaby, British Columbia. In the mid 1990s, Ballard developed a fuel cell that could deliver about 1000 watts per litre of cell, more than six times the previous best fuel cell. Their fuel cells are now in the 1300 watts per litre range and this makes them practical for road vehicles. Indeed, large automakers, such as Ford and Daimler Chrysler have made business alliances with Ballard and several fuel-cell-powered cars and buses have been tested and demonstrated (see Figure 5-3).

Ballard Power Systems

Figure 5-3: An XCELLSIS bus powered by its pre-commercial zero-emission fuel cell bus engine, which uses Ballard fuel cells. This bus was designed using data gained during in-service testing of prototype fuel cell buses in Chicago and Vancouver.

If hydrogen or hydrogen-powered fuel cells are to replace gasoline in the future as a primary transportation fuel, very large amounts of hydrogen will be required. In addition to electrolysis, hydrogen can be generated by a process known as "steam reforming" that removes hydrogen from molecules containing it. The best molecule to use is methane (CH_4), the main constituent of natural gas. In this method, natural gas could be carried in a tank on a vehicle and "reformed" as needed to supply hydrogen fuel. Reforming does produce carbon dioxide, a green-house gas. Nevertheless, fuel cells would produce about half the amount of carbon dioxide as burning the natural gas. Other possibilities include making hydrogen from gasoline or from methanol (the alcohol related to methane).

Every indication is that fuel cells will at some point replace traditional gasoline engines. Building the new infrastructure for the hydrogen fuel cell economy will need huge investments. The rate at which this will occur depends on a variety of economic factors including the premium people are willing to pay for environmental sustainability.

Energy Self Sufficiency
Another important consideration in selecting an energy source is national energy independence. In 1989, the United States spent over $40 billion to import 50% of its oil, which accounted for about 40% of the US trade deficit. This sum grew to about $100 billion at the turn of the century and, given President Bush's aggressive energy policy, will likely continue to grow rapidly in the new millennium. It is small wonder that the US is quick to get involved when there are crises in major oil-producing regions.

It was national energy independence that caused France to make a major commitment to nuclear power after the 1973 oil crisis.

Economics
The bottom line in selecting amongst energy sources is cost. Because of its complexity, the capital cost of constructing a nuclear power plant is higher than for fossil-fuel power plants and, thus, is more sensitive to factors such as licensing, construction time, and interest rates. Some of these factors may be influenced by public opinion, which can cause delays and inflated costs that other energy sources may not have to face. It may also influence consumer choice in an open market. The complexity of a nuclear plant also means it must be carefully managed by a larger and more highly-trained work force to avoid expensive downtime. The higher initial capital cost is compensated for by lower fuelling costs during the ongoing operation of the plant.

To decide whether nuclear or fossil-fuel plants are the best choice economically requires assumptions about future interest rates and fuel costs. Therefore, meaningful comparisons are sometimes difficult to make. For example, the year 2000 saw large (60%) increases in the price of natural gas in North America. Throughout most of the preceding decade generating electricity with natural gas was economically attractive. Ontario Power Generation had proposed to convert one of its coal-fired stations to natural gas for environmental reasons, but the escalating gas prices caused this plan to be dropped.

A new power reactor of the Darlington type would cost around $2 billion to construct. Sums such as this are indeed overwhelming and are perhaps best viewed by placing them in perspective with the costs of other energy projects. Newfoundland's Hibernia oil field, for example, has cost billions of dollars to develop, including investment in advanced technology for platforms as well as drilling and extraction methods. The Hibernia oilfield is estimated to contain 525 to 650 million barrels of oil which can produce 3 to 3.5 billion gigajoules of thermal energy. The eight CANDU reactors at Bruce running at 80% capacity will generate the equivalent amount of electrical energy in 25 years. The Canadian government has invested $3.5 billion in Hibernia since 1979. The total investment in the nuclear research and development program by the federal government since 1945 has been about $5 billion, of which only a small fraction was for research and development directly related to the Bruce nuclear reactors.

In summary, to meet growing electrical demand most countries, including Canada, will not be able to rely on any one energy source but will need to use a mix of sources whose choice depends on many logistical factors, including cost. Nuclear is but one piece of the energy supply puzzle.

SAFETY ASPECTS OF ENERGY

In this section we consider the comparative safety aspects of different energy sources. Detailed discussions of nuclear safety and the impact of nuclear power generation on the environment are discussed in Chapters 8 and 9, respectively.

The public, usually with the Chernobyl accident in mind, views nuclear energy as being particularly dangerous. Detailed assessments of risk, however, demonstrate that nuclear is actually one of the safer sources of energy, even when reactor accidents are considered.

THE HINDENBERG

In the same way that every discussion of nuclear power must mention Chernobyl, the Hindenberg disaster is always raised in the context of a future hydrogen economy. The Hindenberg was a German airship, basically a rigid structure filled with hydrogen gas and propelled through the air by aircraft engines. Containing 190 million litres of hydrogen gas in its 245 metre length it was the largest object ever to fly. On landing after one of its regular trans-Atlantic voyages in May 1937, it caught fire killing 36 people. The film of the disaster is a classic of early news reporting and has been widely shown ever since, sometimes to illustrate the dangers of hydrogen. Recent research, however, suggests that the fabric used in the Hindenberg was easily ignited by atmospheric electricity and was more likely to have caused the fire than the hydrogen.

The potential for severe accidents exists for all major energy systems. Table 5-3 shows deaths caused by different energy sources in the years 1969 to 1986. Hydro-electric power has caused significantly more immediate deaths than any other energy source per unit of energy generated. This perhaps surprising fact is due to the devastating effect of dam failures. For example, some 2,500 people perished in a single dam failure in Macchu, India in 1979.

The main safety concern with coal is in mining, one of the most hazardous occupations in the world. Coal mining is ten times more dangerous than uranium mining (per unit of energy) in general, and probably 100 times more dangerous than for Canadian uranium mining in Saskatchewan. Methane gas and coal dust can explode with devastating results. For example, there were 361 deaths at Monongah, West Virginia, December 6, 1907. Canada has suffered its share of devastating coal mine explosions, the most recent claiming 26 lives on May 9, 1992, at the Westray Mine in Nova Scotia. In 1983 there were 261 uncontrolled coal fires in the USA. Coal mining kills about 100 people per year in the US, down significantly from about 1,000 per year in the early 1900s. There are also many deaths from occupational exposure to coal dust, namely, the respiratory disease known as "black lung".

Table 5-3 shows that natural gas has caused several hundred deaths, generally by explosions from gas leaks. A gas pipeline explosion in Guadalajara, Mexico in 1992, for example, killed 200 people.

Table 5-3
IMMEDIATE FATALITIES FOR SEVERE ACCIDENTS
FROM DIFFERENT ENERGY SOURCES (1969-1986)

Energy Option	No. of Events	Fatalities	Fatalities/Unit Energy
Hydro	8	3,839	1.41*
Coal	62	3,600	0.34
Natural Gas	24	1,440	0.17
Oil	63	2,070	0.10
Nuclear	1	31	0.03

* fatalities per 1000 megawatt-years

Fatalities in the oil sector are from oil platforms capsizing, refinery fires, and fires/explosions during transportation. In 1982, the semi-submersible drilling rig Ocean Ranger capsized and sank on the Grand Banks of Newfoundland, 170 miles east of St. Johns. The entire crew of 84 died, 69 of whom were Canadian. This was Canada's worst oil-related accident.

The table also shows that there were 158 severe energy-related accidents during the period 1969 to 1986. Only one of these was a nuclear accident, although it was certainly a severe and highly publicized one: the Chernobyl accident, which claimed 31 lives. On the basis of immediate deaths, nuclear clearly has the best safety record for accidents of any of the major energy sources.

It should be noted that delayed fatalities, such as possible late cancer deaths from the Chernobyl nuclear accident and lung cancers from air emissions from coal and oil-fired power plants, were not included in Table 5-3 due to the difficulty in obtaining reliable data. It is recognized, however, that coal has the worst impact in this area. Coal-fired electrical generation is a major contributor to the degradation of Ontario's air quality, which causes more than one thousand deaths per year. Oil also has significant delayed health impacts and contributes to air quality degradation through air emissions from motor vehicles.

A SUSTAINABLE ENVIRONMENT

More and more, the energy debate is turning away from economics and focussing on environmental impact. There is a growing realization that the rapidly expanding population is placing an enormous

AN ENVIRONMENTAL COMPARISON OF COAL
AND NUCLEAR POWER PLANTS (1,000 MW EACH)

	Land Use (hectares)	Fuel Use (tonnes/year)	Wastes Generated (tonnes/year)
Coal	70	3,000,000	Ash: 750,000 CO_2: 7,000,000 SO_2: 900 NO_x: 4,500
Nuclear	20	50	900

A 1,000 MW coal station needs more than three times as much land as a comparable nuclear plant, primarily to store coal. In one year, the coal station burns approximately three million tonnes of coal, compared to 50 tonnes of uranium fuel (60,000 times less). A coal-fired station with modern pollution-control equipment creates over 8,000 times more waste than a nuclear plant, most of which is dispersed into the atmosphere; nuclear waste is carefully contained.

stress on the environment, as evidenced by the hole in the atmospheric ozone layer, global warming, air pollution, loss of species, and deforestation, to name just a few of the symptoms. Our voracious demand for energy has been a significant contributor to these problems. Thus, there is an urgent need to integrate health and environmental considerations into the growing electrical demand; we must strive to produce electricity in a manner that supports sustainable development.

In the 1970s, global environmental degradation began to be a serious issue. This concern was first articulated in the United Nations Conference on Human Environment held in Stockholm in 1972. This Conference established the United Nations Environment Program with the responsibility of building environmental awareness and stewardship. The independent World Commission on Environment and Development was also established with the mandate to look at how development affects the environment.

The Commission report, *Our Common Future*, 1987, contained the first reference to the term "sustainable development", which soon captured the world's imagination. The report offered the following definition: "Sustainable development is development that meets the needs of the present without compromising the ability of future generations to meet their own needs."

In 1988, over fifty world leaders supported the report. In 1992 the United Nations Conference on Environment and Development, known as the Earth Summit, was held in Rio de Janeiro and provided enormous publicity and support for sustainable development and produced many initiatives to implement this concept. Sustainable development has become the cornerstone of many government policies.

Natural gas and oil are relatively scarce commodities that are irreplaceable once consumed. Although the world weathered the oil crisis of 1973 and supplies seem adequate at the current time, they will dwindle over the coming decades. Oil and gas are convenient sources of energy, and ecologically better than burning coal, but they should be preserved and used for the functions that they serve best, such as fuel for transportation and feedstock for the petrochemical industry. How do we explain to future generations that we ravenously consumed these resources?

Of the fossil fuels, only coal, which is formed from plants that were buried underground and then subjected to intense heat and pressure over millions of years, is found in such great abundance that it will be available for several centuries. Since 1950, the use of coal has more than doubled and today it accounts for 27% of the world's commercial energy. Coal is used to produce 39% of the world's electrical energy and 75% of its steel. The downside is that coal is environmentally the "dirtiest" fuel.

Table 5-4	LIFE-CYCLE EMISSION OF CARBON DIOXIDE FOR ELECTRICAL GENERATION SOURCES	
	Energy Source	CO_2 (grams/kilowatt-hour)
	Coal	about 1000
	Solar	60-150
	Nuclear	6
	Wind	3-22

The generation of green-house gases that create global warming has become an international issue. Table 5-4 summarizes the capacity of four different electrical-energy sources to contribute the greenhouse gas, carbon dioxide, to the atmosphere on a unit-of-energy-produced basis calculated over a full life-cycle, including manufacturing, installation, and operation of the facilities. As expected, coal creates the most green-house gases per unit of energy. Perhaps surprisingly, nuclear is comparable to the renewables, wind and solar. These require the manufacturing of many units, due to the diffuse nature of solar and wind energy, and this manufacturing process uses energy, which releases carbon dioxide.

Even hydro, a renewable and seemingly benign energy source that creates no air emissions, is not without problems. The construction of dams and reservoirs requires the flooding of large areas of land, displacing many inhabitants and disrupting wildlife and fish habitats as well as historically significant sites. The James Bay project in northern Quebec is a good example. As originally envisaged, the 50-year megaproject was to build 600 dams and dikes, flood an area of 176,000 square kilometres, and displace thousands of indigenous Cree and Inuit. After 20 years and $16 billion, Phase 1 was completed with the flooding of 11,000 square kilometres of boreal forest and tundra. Faced with strong opposition and a surplus of power, the second phase was postponed indefinitely in 1994.

The Aswan hydroelectric dam in Egypt has caused ecological problems. The vast Three Gorges project in China may be another. This hydro power megaproject will displace millions of people, flood important historic and archaeological sites, and is facing formidable opposition from local populations.

When dams flood large areas of land the underlying vegetation decays producing methane. Since methane is 20 times more damaging as a greenhouse gas than carbon dioxide, some dams involving the submerging of large masses of trees and plants may cause more global warming than they avoid by not having to burn fossil fuels to generate the same amount of electricity. The same decay processes release a soluble form of mercury, methyl mercury, from the rocks and soils that is toxic to humans and wildlife.

Although uranium is a finite, non-renewable resource, nuclear reactors can be modified to use other fuels such as thorium, which is more abundant, and also to generate, or "breed", new fuel. Nuclear technology has the potential to be sustainable, that is, to provide an essentially infinite supply of energy. Although this may seem like an utopian dream, technologies currently exist that allow more nuclear fuel to be created than is consumed; these are described in Chapter 14.

THE ROLE OF NUCLEAR ENERGY

Nuclear is a different kind of energy; it is not produced by burning a fuel and thus does not produce atmospheric emissions. Under normal operations, nuclear plants are environmentally clean, and their performance under accident conditions (as shown in Table 5-3) is better than that of other energy sources.

For these reasons, many organizations including the Club of Rome and the Union of Concerned Scientists have supported the use of nuclear power to replace increased burning of fossil fuels.

In 1988, the Canadian Parliamentary Standing Committee on Energy, Mines and Resources, concluded that "The environmental impacts of burning larger amounts of fossil fuel, especially coal, to generate electricity has become alarming. Research is revealing the magnitude of the public health hazard, the enormous economic costs and the environmental destruction resulting from acid gas emissions from fossil-fuelled power plants. The implications of carbon dioxide accumulation in the Earth's atmosphere — an unavoidable accompaniment to fossil fuel combustion — are being studied intensively and the potential for disruptive climatic change is evident. Set against these concerns, the Committee finds nuclear power to be an environmentally appealing technology." These comments were made by exhaustive research and testimony, including presentations by anti-nuclear groups.

As we enter the new millennium, we are faced with some difficult challenges. The world's growing population desperately needs energy. In addition, the global environment is suffering. Nuclear energy, if properly used, can make an important contribution to solving these problems.

Chapter Six

CANDU:
The Canadian Reactor

In 1987, the CANDU reactor, along with the CN Tower and the Alouette space satellite, was ranked as one of Canada's top ten engineering achievements during the previous 100 years. In making the award it was stated that, "Perhaps the greatest challenge posed by CANDU was the necessity of adhering to extremely high quality standards in every aspect of the project – in design, manufacture of components, construction and maintenance and operation. This resulted in advances and activity in numerous other related fields. Perhaps best known are the medical applications, such as cobalt therapy machines for cancer treatment. Other spinoffs from the nuclear program include automatic computerized control systems, simulator models, remote handling techniques, and fundamental advances in areas such as metallurgy and chemistry."

The CANDU reactor has clearly had a significant impact on Canadian technology; let us look at it more closely. A total of 22 CANDU power reactors have been constructed and operated in Canada. CANDU stands for Canada Deuterium Uranium, indicating the use of deuterium, or heavy water, for moderator and **natural uranium** for fuel. Typical CANDU reactors are shown in Figures 2-5 and 2-6.

REACTOR BASICS

Although nuclear reactors are complex, the basic principles under-lying their operation can be understood relatively easily. A nuclear chain reaction is used to produce heat. The heat turns a fluid to steam and this steam is used to turn turbines and generate electricity. A schematic of the CANDU reactor illustrating these features is shown in Figure 6-1. The basic concept is similar to that of a coal or oil-fired power station, except that the energy source is nuclear fuel. The main difference is in how the heat is produced. In a nuclear reactor the heat comes from splitting the uranium-235 nucleus; in a fossil-fuel power plant, the heat comes from burning coal, oil, or natural-gas.

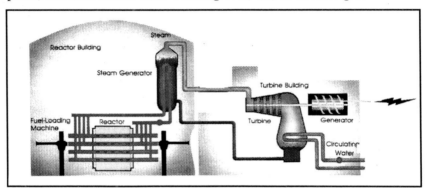

Figure 6-1: Schematic of a CANDU reactor. Pressure tubes containing fuel bundles pass horizontally through a vessel called a calandria containing heavy-water moderator. A nuclear reaction in the fuel bundles heats the heavy-water coolant; this heat is carried to the steam generator where it is transferred to ordinary water in a separate circuit to produce steam. The steam is sent through a turbine, which turns a generator to produce electricity. Circulating water from outside condenses the steam to water, which is then cycled back to the steam generator. The figure also shows the fueling machines and the reactor contain-ment building.

Nuclear reactors have the following main components.

Fuel: First, a nuclear reactor needs a **core** of fuel whose nuclei will fission or split when hit by a neutron. This fissile material must be in a sufficiently high concentration that it will sustain an ongoing chain reaction. Virtually all reactors in the world use uranium as the fuel, which in nature is composed of two isotopes, uranium-235 (0.7%) and uranium-238 (99.3%). A significant complication is that only the less abundant of these, uranium-235, can be fissioned. As an aside, there are other fissile materials such as plutonium-239 and uranium-233 but these are not currently used to make nuclear fuel in North America (this topic is discussed in Chapter 14).

There are two fundamental approaches to achieving a nuclear reaction in a reactor. One is to increase the concentration of the fissile uranium-235 fuel to the point where a self-sustaining chain reaction is possible using ordinary water as moderator. This is the approach adopted by most of the reactors in the world, called light-water reactors, which use uranium-235 fuel **enriched** to 3% to 5% (these reactor types are described in Chapter 7). The approach in the CANDU is to use a far more efficient moderator, heavy water, which allows the use of natural uranium (without enrichment) as a fuel.

Moderator: The neutrons emitted when the uranium-235 nucleus fissions are travelling at such a high velocity that they speed right past other uranium-235 nuclei without splitting them. Thus, the core of fissile fuel must be surrounded by a moderator that slows the neutrons to an appropriate speed, yet does not absorb them. Only a few materials, such as heavy water, graphite, and beryllium are capable of doing this efficiently, that is, without absorbing neutrons. Ordinary water can also be used, but the concentration of uranium-235 must be increased to allow for neutron absorption. Moderated reactors are known as **"thermal" reactors** because fission occurs with slow neutrons, that is, the neutrons have very low kinetic energy and are known as thermal neutrons.

Coolant: The enormous amount of heat that is generated in the core must be removed to drive turbines and prevent melting of the fuel and other damage. Fluids that are commonly used include both ordinary and heavy water, organic liquids, liquid metals (sodium), and gases such as carbon dioxide and helium.

Control Systems: The rate of nuclear fission must be precisely controlled to maintain the chain reaction at the proper level: too low a rate of fission and the chain reaction will terminate; too high a rate and more heat will be generated than the coolant can remove. Typically, rods of stainless steel or alloys containing boron, which is particularly good at absorbing neutrons, are moved up and down inside the core to regulate the reaction. Rods containing light water can also be used.

Safety Systems: Because of the large amount of energy that is involved, it is essential that a nuclear reactor have a very reliable and fast-responding shutdown system. One of the most widely used methods to stop the nuclear reaction is to insert rods made of cadmium, a strong neutron absorber, into the core. Alternatively, gadolinium, another strong neutron absorber, can be injected in liquid form into the moderator. As a safety measure, reactors are generally equipped with independent control and safety systems.

Steam Generator: The heat in the coolant is transferred via a heat exchanger to an independent secondary circuit where boils and generates steam to drive turbines. Normally the primary circuit is kept under sufficient pressure to prevent the water from boiling, whereas the water in the secondary circuit is maintained at a lower pressure so it boils. Steam generators, large vessels filled with hundreds of tubes, pass the heat from the coolant circuit to water in the secondary circuit. This system has the advantage that it keeps all the radioactivity in the primary circuit, isolated from the turbines. The steam generators serve as heat exchangers and are sometimes called boilers. In boiling water reactors, the cooling water is allowed to boil and the resultant steam is sent directly to the turbines (see Chapter 7).

Turbine/Generator: Steam turns a turbine which causes a generator to spin and create electricity. Steam generators and turbine/generators are the same as in conventional power plants fuelled by coal, oil, or natural gas. In both nuclear and non-nuclear power plants about two-thirds of the heat still remains in the steam after it has passed through the turbine. This is removed to condense the steam back to water, and is dissipated as waste heat either to a body of water or to the atmosphere using cooling towers.

CANDU REACTOR DESIGN

The CANDU has several unique design features that distinguish it from other international reactors.

Fuel: The CANDU is fuelled by natural uranium with a concentration of 0.7% of the fissile uranium-235. This avoids the complex and costly process of enriching the uranium-235 concentration. Although uranium-235 provides the main fissile fuel in a CANDU, a small amount of uranium-238 is converted by neutron absorption to uranium-239, a fissile isotope that subsequently fissions and contributes about 40% of the heat. A CANDU fuel bundle is shown in Figure 6-2. Zirconium metal is used to contain the uranium since it absorbs far fewer neutrons than steel. A Pickering fuel bundle contains 28 elements whereas a Bruce or Darlington bundle contains 37 elements. A new design of fuel bundle with 43 elements has been developed that allows more efficient heat removal. (See Chapter 14)

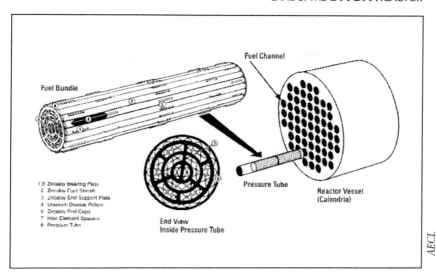

Figure 6-2: Schematic of a CANDU fuel bundle consisting of a number of zircaloy tubes filled with pellets of uranium oxide. The bundles are inserted into horizontal pressure tubes passing through the reactor vessel.

Moderator: Because natural, rather than enriched, uranium is used in a CANDU, a chain reaction cannot be sustained by using ordinary water as a moderator. Instead, a much more effective moderator, heavy water, that is, deuterium oxide, is used. Since heavy water has almost all the extra neutrons it wants, it slows neutrons in the reactor without significantly absorbing them. It is 1700 times more efficient as a moderator than ordinary water. When a neutron is emitted by a splitting nucleus, it travels at high speed through the fuel and the fuel **cladding**, passing into the heavy-water moderator where it is slowed down to 'thermal' speed. It then passes into another fuel channel and strikes a uranium-235 nucleus, causing it to split.

The production of heavy water is complex and costly, adding significantly to the initial capital cost of a CANDU (this initial cost is later re-captured by cheaper ongoing fuel costs). The more recent CANDU reactors contain about 0.8 tonnes of heavy water per megawatt, with an annual loss of about 1%.

Reactor Configuration: Fuel is arranged in horizontal pressure tubes inside a vessel called a **calandria**. Because of the lower power density in a CANDU core, due to the natural uranium fuel, the reactor core is appreciably larger than for light-water reactors of similar capacity. Because of this size, the early CANDU designers decided to use a number of pressure tubes rather than a single large pressure vessel to contain the fuel. Although only four pressure tubes are shown in Figure 6-1, there are actually several hundred in a CANDU reactor. The heavy-water coolant flows through these pressure tubes, each of which is enclosed inside a calandria tube. The two tubes are separated by spacers called garter springs. This arrangement keeps the coolant and moderator separate. The space between the two tubes is filled with carbon dioxide gas, which acts as a thermal insulator and is monitored to detect leaks.

The calandria contains several hundred tonnes of heavy-water moderator under low pressure. To ensure a high chemical quality, the moderator is continuously circulated through purification systems. It is kept at a relatively low temperature (approximately 70°C) using a separate heat-exchanger cooling system.

Because the moderator is under constant neutron bombardment, over time some of the deuterium absorbs neutrons to form tritium. As tritium is radioactive, it must be removed from the moderator periodically.

There are about 380 pressure tubes in a typical CANDU 6 reactor such as at Point Lepreau, New Brunswick, and each tube contains 12 fuel bundles, so that there is a total of 4,560 bundles in the reactor core. Larger reactors like Darlington have more fuel channels (480) and thus more fuel bundles per channel. A bundle stays in the reactor between six and 24 months, depending on its location in the core.

The reactor calandria is contained in a thick-walled concrete and steel structure that provides shielding. In stations built after Pickering A, this shielding structure is filled with hundreds of tonnes of ordinary water and steel balls to provide additional shielding.

The CANDU control and safety systems are described in the next section and in Chapter 8.

In summary, in a CANDU reactor the heat-transport system circulates heavy-water coolant through the fuel channels to remove heat. The coolant, which is kept under pressure to prevent boiling, is then circulated by pumps to a set of steam generators. These act as heat exchangers, transferring heat to a separate ordinary-water system. The ordinary water is maintained at lower pressure so it boils, forming steam which, in turn, drives the turbines to produce electricity. After

W. B. Lewis (1908-1987) was a born in England and was a student of Rutherford at the University of Cambridge. He was invited to come to Chalk River as Director of Research in 1946. He was one of the first scientists to recognize the potential of nuclear energy and championed it in Canada and abroad throughout his lifetime. Lewis was the driving force behind the CANDU reactor and is called "the father of the CANDU". He received numerous honours during his life including the Order of Canada.

AECL

driving the turbine blades, the steam is condensed and cooled using nearby lake, river, or ocean water and then recirculated.

On-Power Refuelling: CANDU reactors have on-power refuelling. This is an important and unique feature, allowing them to achieve high capacity factors since they do not need extensive shutdowns every one or two years for refuelling, as is the case for light-water reactors. This feature has resulted in two other unique design features. First, the fuel channels are horizontal rather than vertical, to allow the fuelling machines to insert fuel bundles at one end of the core while extracting them at the opposite end. A photo of a refuelling machine at a calandria face is shown in Figure 6-3.

Second, fuel bundles can be moved in the fuel channels, usually in two or three stages, to achieve maximum **burnup**. Combined with the very efficient moderation provided by heavy water, this results in CANDU reactors using about 25-30% less natural uranium over a 30-year lifetime than a comparable light-water reactor.

The CANDU pressure-tube design, although offering benefits, also has some shortcomings. While the capital costs of CANDUs are comparable to those of their competition, they involve a lot of "plumbing". As each fuel channel is contained in its own pressure tube, it makes a larger and more complicated core than for light-water reactors, contributing to the higher cost of maintaining CANDU reactors.

The amount of engineering required to develop a CANDU reactor is enormous. The refuelling machines alone are extremely complex machines that must operate remotely in extreme temperature and radiation fields. In addition, there are numerous pumps, valves, temperature and radiation sensors, control systems, and many other complex components.

Ontario Power Generation

Figure 6-3: Fueling machine at the face of a CANDU reactor. The machine attaches to the end a fuel channel and adds or removes fuel bundles while the reactor is operating, a unique feature of the CANDU reactor.

CANDU OPERATION

How do reactor operators control such a large and complex piece of machinery? How do they start up, control, and shut down the nuclear fission process, particularly in a manner that ensures safe operation?

Parameters such as neutron flux, coolant flow, and temperature are measured at a number of key points, usually in duplicate or triplicate. These are monitored by computers and by the reactor operators in the control room. If any measurement is outside its specified range, the power is automatically adjusted either in the particular zone or in the whole core. Four devices are used to control the number of neutrons and hence the power of the reactor: light-water zone controls, **control rods**, adjuster rods, and moderator **poison**.

Light-water zone controllers are tubes running vertically through the core into which light water can be introduced. The light water absorbs neutrons and lowers the power in that zone. The calandria volume is divided into 14 "zones" in each of which the power is controlled by a zone controller.

Control rods are held vertically over the reactor core and can be lowered into the core at varying speeds. These rods contain cadmium, a very good neutron absorber, in a stainless steel sheath. The rods are mechanically controlled by computer but can also be operated manually. Depending on the speed of insertion, they can be used to slowly change reactor power or to quickly shut down the reactor in an emergency. An advantage of the CANDU system is that the shutdown rods are inserted into the calandria against little or no resistance as the moderator is not at high pressure, as is the case for most other reactor types.

Adjuster rods are usually fully inserted into the reactor core to absorb neutrons, with more absorption in the central region, thus keeping the power relatively constant across the core (this is called flux shaping). They can also be withdrawn as fuel gets older to compensate for the build up of fission products, which absorb neutrons. The adjuster rods can also be withdrawn to allow for short-term inability to refuel. Although usually made of stainless steel, in some reactors the rods contain cobalt-59, which upon irradiation becomes cobalt-60. The cobalt adjuster rods are periodically removed from the reactor and sold to MDS Nordion where the cobalt-60 is removed and repackaged for industrial uses such as sterilization. About 85% of the cobalt-60 used in the world comes from Canadian reactors.

Another important method of ensuring a uniform distribution of neutrons is by arranging the fuel bundles using the on-line fuelling machines. New fuel bundles can be placed in areas where fuel bundles have been in the reactor longer to ensure a level neutron distribution throughout the reactor core. Continuing adjustments can be made as the fuel is used up.

Another independent shut-down system uses the rapid injection of liquid gadolinium nitrate, a strong neutron absorber, into the moderator. Although generally reserved for emergencies, this method can also be used for routine control. The "poison", subsequently needs to be removed from the moderator by passing it through ion-exchange resins.

In earlier CANDU reactors it was possible to dump all of the moderator out of the calandria into a holding tank. This would immediately stop the chain reaction, but it is now recognized that, in an accident scenario, the moderator should remain in place to help absorb heat from the fuel. Thus, this means of shutdown is no longer used.

CANDU OPERATING PERFORMANCE

Until recently, CANDU reactors have had excellent performance records thanks largely to their unique on-line refuelling capability. In 1988, for example, CANDU reactors, with gross capacities ranging from 84.9 to 90.3 percent, took seven of the top nine places for life-time performance among all reactors in the world over 500 MW in capacity. In 1987, CANDU reactors had probably the best showing ever by any reactor type, holding the top six as well as ninth and twelfth positions amongst world reactors for life-time capacity. Inaddition, Pickering Unit 7 established a world record for the longest non-stop operation by any nuclear power reactor of 894 days.

In August 1983, however, the CANDU reactor suffered a major technical setback that would significantly reduce overall performance in future years. While on full power, a pressure tube ruptured at Pickering Unit 2, spilling mildly radioactive heavy water from the primary circuit into the reactor building. Investigations revealed that the spacer (or garter spring) that separates the pressure and calandria tubes had migrated from its position due to vibration, causing the pressure tube to sag and touch the colder calandria tube. This led to embrittlement of the zirconium-tin alloy and finally a fracture. There was no release of radioactivity to the environment and the reactor was safely shut down by the operators before the emergency safety systems were required.

It was soon discovered that all four units at Station A would require retubing. The decision was made to retube Units 1 and 2 at the same time. Unit 2 was out of service for 62 months while all of its pressure tubes were replaced; the retubing of Unit 1 was done in 47 months. The cost was about $400 million, and further costs were incurred in providing replacement energy during these lengthy outages. With the experience gained, the time and cost for retubing Units 3 and 4 decreased significantly, taking 26 and 19 months, respectively.

It was always recognized that CANDU reactors would require retubing at some point during their lifetimes, seriously diminishing the advantage of on-line refuelling. The very short lifetime of the Pickering Station A pressure tubes (about 12 years) was a cause of great concern at that time. Intensive scientific and engineering effort was and continues to be directed at this problem. Methods have been developed for regularly monitoring the condition of pressure tubes and repositioning garter springs so that retubing should not be

Harold Smith (1919-2000) was born in Lucknow, Ontario, and graduated from Queens University in electrical engineering. After war service in the Royal Canadian Navy, he joined Ontario Hydro where he rose to become leader of the Nuclear Power Group. He was a prime mover in developing the Canadian nuclear power program and is credited with inventing the name "CANDU". Under his technical leadership Ontario Hydro became one the world's largest nuclear utilities.

Ontario Power Generation

required until after about 25 years of operation. In reactors constructed since the early 1990s, new materials (zirconium-niobium alloy with much lower hydrogen content and lower corrosion) as well as a revamped design of the garter spring ensure that retubing should not be required for about 30 or more years, and then not all tubes may need replacement.

Research continues to make future retubings less frequent and more predictable with shorter reactor downtime to conduct them.

The Ontario Power Generation Saga

Ontario Power Generation (formerly called Ontario Hydro) has often been in the glare of the political spotlight, but at no time has it been more blinding than in recent years. Ontario Power Generation went through a painful and deep-cutting corporate restructuring in the mid-1990s. This was followed by a stunning announcement on August 13, 1997, that seven of its 20 reactors would be laid up over the next year.

The decision was based on a comprehensive study that concluded that although CANDU technology was sound, Ontario Power Generation had not managed and maintained the reactors to achieve efficient and high-capacity operation. The downsizing in the mid-1990s and its enormous debt had left Ontario Power Generation with an inadequate workforce and financial resources. Thus, the seven oldest reactors were laid up so efforts could be focussed on developing improved practices for the remaining reactors. The intention was to bring the laid-up reactors back on line in the future. By the beginning of 2001, four of the seven laid up reactors (Pickering Station A) had passed an environmental screening process and the next stage of their return to service plan had been initiated. In early 2001, Bruce Power concluded it was economicaly feasible to bring two of the (Bruce A) reactors back to service.

Ironically, while Ontario Power Generation was stumbling, reactor performance was improving everywhere else in the world. Driven by competition with coal and gas-fired power, nuclear maintenance has received high priority and has resulted in improved international reactor performance.

So what went wrong? Many reasons have been put forward. One of the most prevalent is that it had been a mistake to go nuclear in the first place. This is not necessarily the case, as in other jurisdictions such as Quebec, Argentina, Romania, and Korea, CANDU reactors have performed well.

It has also been said that Ontario Power Generation built too much capacity. This is true, but who was to know in the 1960s and 1970s that a long period of high electrical growth, in which demand doubled every ten years, was to grind to a snail's pace by the early 1990s. Every other province also committed to excess capacity during this period.

The root cause appears to lie in Ontario Power Generation operating as a centralized, government-controlled monopoly. It was not responsible to shareholders and lacked the checks and balances that should have protected its capital investment, the reactors, from deteriorating. Recognizing this problem, Ontario Hydro was split into three separate companies in 1999 and its monopoly over electricity generation in Ontario was ended. The new power-generating entity, Ontario Power Generation, is to operate in an increasingly competitive market place and eventually will only control a maximum of 35% of electrical generation in Ontario.

ECONOMIC IMPACTS

The Canadian nuclear industry is far larger than most people realize and makes a significant impact on the nation's overall economy. Some economic highlights are summarized below:

• CANDU reactors in Canada produced energy valued at $3.7 billion in 1992.

• There are 30,000 direct and 10,000 indirect jobs in the Canadian nuclear industry; many of these are high-technology jobs

• The use of uranium to generate energy saved $1 billion in foreign exchange costs for importing coal in 1992.

• The nuclear industry exports more than it imports. The only other high-tech industry to have an export surplus was the aerospace industry. All the others such as biotechnology, weapons, computers, electronics had negative export balances.

• Canada is the world's leading exporter of uranium, earning more than $1 billion per year.

The above data was produced by Ernst & Young (1993), one of the top accounting firms in Canada, using established accounting methods. They concluded that "the economic benefits of the Canadian nuclear industry have been substantial. Over the past 31 years, the GDP contributions of the nuclear power generation industry has been at least $23 billion (as-spent dollars). The GDP contributions for 1992 were $3.5 billion."

Many high-technology companies have sprouted from the Canadian nuclear industry, including MDS Nordion, Zircatec Precision Industries, GE Nuclear, Spar Aerospace, Cameco, Babcock and Wilcox, and CAE Electronics.

Quality assurance and quality control standards used in the nuclear industry are among the highest in any industry. Canadian manufacturers and suppliers to the nuclear program have had to adopt and work to these stringent standards, which have been applied to other product lines and have strengthened the competitive position of Canadian firms in the international arena.

THE CANDU OVERSEAS

AECL has been successful in selling the CANDU reactor internationally and has held its own competing against such industrial giants as Westinghouse, General Electric, France's Framatome, and Germany's Siemens and Kraftwerk Union.

The CANDU's international involvement began in the early days of Canadian reactor development. In 1955, India and Canada opened

85

discussions which led to Canada supplying an NRX-type research reactor, called CIRUS, to India under the Colombo Plan, a Commonwealth program for support to underdeveloped countries. In the late 1960s, two Douglas-Point type reactors were sold to India, called RAPP-1 and RAPP-2.

The 137 MW KANUPP reactor was constructed near Karachi, Pakistan in 1972 by General Electric Canada. The KANUPP was modelled after the Douglas Point reactor. It continues to operate today as Pakistan's only reactor (although a Chinese-design reactor is currently under construction) and was the only overseas CANDU reactor not sold by AECL.

In 1974, after RAPP-1 was on line, but before the completion of RAPP-2, India detonated a nuclear bomb. Canada immediately stopped all nuclear assistance to India. Without any Canadian assistance, India went on to complete RAPP-2 and develop its own indigenous nuclear industry, based on the pressurized heavy-water reactors supplied by Canada.

A CANDU reactor was sold to Argentina and began commercial operation at Cordoba in 1984; it is known as the Embalse reactor. Embalse has maintained a lifetime average of 80.6% capacity factor. Together with Atucha-1, Argentina's other nuclear power reactor, they supply about 12% of the country's electricity requirements.

THE INDIAN "CANDUS"

India has an expanding domestic nuclear power program based on a pressure-tube, heavy-water reactor design provided by Canada in the early 1970s. Originating from the RAPP-1 project, built in the same era as the Canadian Douglas Point prototype, they are small compared to typical CANDUs, the largest being 220 MW. Twelve of these reactors now generate electricity deployed in one 4-unit station at Rajasthan and in four 2-reactor stations elsewhere. They are not called "CANDUs" since this is a registered trade mark of AECL and Canada had no part in their construction because of India's nuclear weapons program. It is interesting to note that the two reactors provided to India by the United States in the 1960s, prior to any Canadian transfer of nuclear technology, are still generating electricity today as the only "non-CANDU" reactors in the Indian power grid.

AECL

Figure 6-4: Four CANDU 6 reactors at Wolsong on the coast of Korea. These reactors have had an outstanding performance record.

Korea has made a major commitment to using nuclear power, with a mix of different reactor types, including the CANDU and light-water reactors. Four CANDUs have been purchased from AECL and installed at Wolsong (see Figure 6-4) ; they came into commercial production in 1983, 1997, 1998, and 1999.

In the late 1980s Romania expressed interest in purchasing four CANDUs to be constructed at Cernavoda. Only Cernavoda-1 proceeded and, due to delays in obtaining financing, it was not connected to the Romanian electrical grid until 1996. The completion of Cernavoda-2 and the construction of the remaining units is currently under discussion.

China purchased two CANDU 6 units in 1997. These are located at Qinshan and are scheduled to come on line in 2003 and 2004.

It is interesting to note that the international CANDU reactors have not encountered the maintenance problems that have recently plagued Ontario reactors. In 1993, for example, the South Korean electric utility revealed that, not only is nuclear its lowest cost source of electricity, but that its CANDU plant was significantly lower in cost than its light-water reactors.

In contrast to the lengthy delays at Ontario's Darlington station (where the four reactors were completed in 9-10 years), Korea was able to construct and bring Wolsong-3 into commercial service in a record four years and nine months from award of contract.

Chapter Seven

The Global
Nuclear Picture

In spite of the controversy that sometimes surrounds it, nuclear power continues to be a key component of today's global energy mix. In 2000, about 19% of the world's total electrical capacity was generated by 433 nuclear power reactors located in 31 countries. These reactors have an electrical generating capacity of approximately 349,000 MW. Another 31 reactors were under construction with an additional 31,000 MW of capacity.

LIGHT-WATER REACTORS

There are two main power reactor types in use in the world: those moderated by **light water** and those moderated by heavy water. The CANDU is a pressurized heavy-water reactor (PHWR). The light-water reactors are divided into two types: the boiling-water reactor (**BWR**) and the pressurized-water reactor (**PWR**). "Pressurized" means the water in the reactor circuit is kept under pressure so it cannot boil and generate steam. Since steam is necessary to drive the turbines, a PWR requires a steam generator to transfer heat to a secondary circuit where water boils to make steam. In the BWR, steam is produced directly in the reactor core.

Light-water reactors (**LWR**s) are the most common type in the world. For example, all of the one hundred or so commercial power reactors in the United States are light-water reactors. (The sole exception, the 330 MW high-temperature gas-cooled reactor at Fort St. Vrain, Colorado, was taken out of service in 1989.) In the USA, boiling water reactors are supplied by General Electric, whereas pressurized-water reactors are supplied by Westinghouse (now owned by BNFL), Combustion Engineering, and Babcock and Wilcox (whose reactor division is now owned by Framatome of France).

The BWR steam cycle employs a single loop as shown in Figure 7-1. In this direct cycle system, the water is heated by the fuel core so it boils and produces steam inside the reactor vessel. The water in a BWR is actually pressurized (to 7.2 Megapascals) to raise the boiling point of water to 288 °C (compared to 100 °C at normal atmospheric pressure); the higher temperature improves the efficiency of electricity generation.

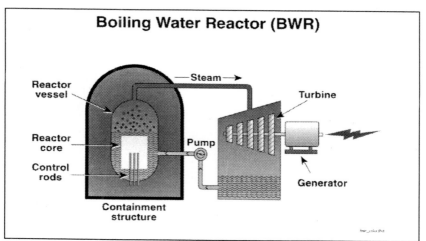

Figure 7-1: Schematic of a boiling-water reactor. Steam is generated in the coolant, which passes through the reactor and is sent directly to the turbine to generate electricity. This makes for a simple design but requires careful control of radioactivity in the coolant. Note that the control rods are inserted from the bottom.

The PWR maintains pressure sufficiently high (15.5 Megapascals) to prevent boiling in the reactor vessel. This type of unit employs a two-loop steam cycle as shown in Figure 7-2. The temperature of the water in the primary loop is about 327 °C as it leaves the core. As in the CANDU reactor, the steam generator transfers heat from the primary to a secondary circuit and produces steam in the secondary circuit for the turbine-generator.

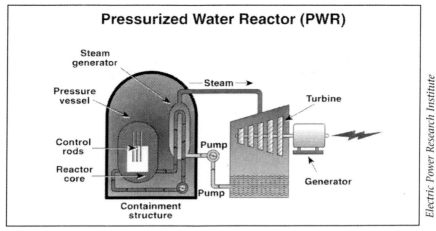

Electric Power Research Institute

Figure 7-2: Schematic of a pressurized-water reactor. As in the CANDU design, the primary coolant loop transports heat from the reactor core to one or more steam generators w hich produce steam in a secondary circuit. The steam is then taken to turbines to generate electricity.

Despite the differences between the PWR and BWR, they have many design similarities resulting from the use of ordinary water as their moderator and coolant. The two are, therefore, grouped together as light-water reactors.

In both the BWR and PWR, the coolant water also serves as the moderator. Because normal water is not as effective a moderator as heavy water, the uranium fuel must be enriched to about 2 to 5% uranium-235. This is a tradeoff. For CANDU reactors, the moderator's isotopic content is modified; in the light-water reactors, the uranium's isotopic content is changed.

Although the BWR is conceptually simpler than the PWR, requiring no steam generators, it allows radioactive coolant water to leave the reactor building and enter the turbine/generator system. Thus, the control of contamination and radiation protection for workers is more complicated. The bottom line is that BWRs and PWRs are comparable in terms of total cost.

The fuel assemblies for the two types of light-water reactor are similar. Slightly enriched uranium dioxide (about 2 to 5% uranium-235) is fabricated into ceramic cylindrical fuel pellets (about 0.8 cm diameter and 1.3 cm long). The pellets are placed into long zirconium-alloy cladding tubes to produce fuel pins, also called fuel rods. The fuel pins for a BWR and PWR are quite similar.

A rectangular array of these pins forms the final fuel assembly. The BWR fuel assembly consists of an 8 x 8 array of fuel pins. A fuel channel encloses the fuel-pin array, so that coolant entering at the bottom of the assembly will remain within this boundary as it flows upward between the fuel pins and removes the heat.

The PWR fuel assembly, which consists typically of a 16 x 16 or 17 x 17 array of fuel pins, is about four times larger than that of the BWR. The array is not enclosed by a fuel channel since the behaviour of the non-boiling coolant is much more predictable than that of the BWR. Typically, a PWR fuel assembly is about 15 cm x 15 cm in cross section and about 4 m long. A generic fuel assembly for a light-water reactor is shown in Figure 7-3.

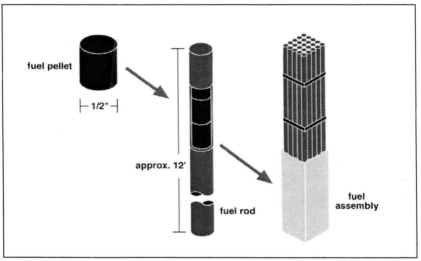

· Electric Power Research Institute

Figure 7-3: A generic light-water reactor fuel assembly. Pellets of enriched uranium are placed into long zirconium-alloy tubes. The rods are bundled together in 16 X 16 or 17 X 17 square arrays to form a fuel assembly. Several hundred such assemblies form the reactor core.

Both the BWR and PWR have vertical reactor vessels. These are large steel structures in which the fuel assemblies, the coolant/moderator water, control rods, and other equipment are contained. In the typical BWR there are 748 fuel assemblies, while there are between 193 and 241 in the typical PWR. The PWR vessel is made of 20-cm-thick steel and is 12 metres high; the BWR reactor vessel is made of 15-cm-thick steel and is typically 21 metres high. The BWR reactor vessel is higher than that of the PWR because it also contains steam separators and steam dryers that remove water from the steam.

In contrast to the CANDU, both the PWR and BWR units must be shut down for refuelling approximately every 18 to 24 months. This refuelling shutdown used to take one to two months but has been reduced considerably in recent years. In 2000, a new US record for the shortest refuelling shutdown, just under 16 days was set by the Braidwood-2 LWR. In addition, the interval between refuelling shutdowns has been increasing.

Many of the BWR and PWR safety systems are similar to those described for the CANDU reactor. Note that PWR control rods are inserted into the core from above, whereas they are inserted into the core from below for the BWR. In both instances, the rods must be inserted into a high pressure zone; in CANDU reactors, control rods are inserted into a calandria kept at low pressure.

The reactor vessels are completely enclosed by a concrete and steel primary containment structure. Both the reactor vessel and the primary containment structures are then contained within a secondary containment structure. In addition to the containment structures, there are numerous other safety systems common to BWRs and PWRs. These systems are intended to prevent or mitigate the consequences of any accident and are based upon `defence by redundancy' as well as `defence in depth' (see Chapter 8).

The principal safety systems utilized in light-water reactors are collectively known as Engineered Safety Systems. These are similar to the CANDU safety systems discussed in the next chapter and have been designed, in particular, to deal with "loss-of-coolant accidents", the worst kind that can happen, and include:
- reactor trip (shutdown) to provide positive and continued shutdown of the nuclear chain reaction
- emergency core cooling to limit fuel melting
- post-accident heat removal to prevent pressure buildup in the containment
- post-accident radioactivity removal to reduce the radionuclide inventory available for release
- containment integrity to limit radionuclide release.

In recent reactors, most of the safety systems are part of the basic design. In older units, many of the safety systems have been added based on lessons learned from the Three Mile Island accident. Figure 7-4 shows the Sequoyah PWRs operated by the Tennessee Valley Authority in the United States.

More recent commercial reactors within the US are generally of larger capacity than their predecessors.

Although no new reactors have been constructed in the USA since 1990, the existing reactors have steadily increased their electrical production. By 1999 electricity equivalent to the output of 19 new reactors has been added to the US electrical grid. Part of this has come from increasing the power ratings of the reactors and the remainder from improved performance. The average capacity factor for US LWRs was 89.7% in 2000.

Table 7-1 summarizes the main differences between the PWR, BWR, and the CANDU reactor. The CANDU core is larger and more complex than the LWR core because of the pressure tubes. However, this provides some extra safety as the pressure tubes separate the moderator from the coolant so that the moderator is kept at a much lower temperature and would help remove heat in case of an accident. In the LWR, the moderator and the coolant are one and the same and are kept at high temperature.

Electric Power Research Institute

Figure 7-4: The Sequoyah nuclear power station operated by the Tennessee Valley Authority. Two 1148 MW PWRs are enclosed in the domed containment structures in the centre of the picture. The two large structures are cooling towers where residual heat is cooled by air flowing in the towers.

Table 7-1 COMPARISON OF BWRS, PWRS, AND CANDU REACTORS

	LWRs	CANDU
Fuel	Enriched Uranium (2-5% U-235)	Natural Uranium (0.7% U-235)
Fuel Assembly	(BWR) 8 x 8 square array (PWR) 16 x 16 / 17 x 17 square both about 4 m long	Bundle of 37 pins in cylindrical arrangement about 10 cm in diameter, 50 cm long
Number	(BWR) 748 fuel assemblies (PWR) 193 to 241 assemblies	4560 to 6240 bundles in core
Reactor Vessel	Large vertical steel tank (pressure vessel)	Calandria (unpressurized horizontal tank) with 380 to 480 fuel channels
Moderator	Ordinary water	Heavy water
Control Rods	Enter into high-pressure, high-temperature zone	Enter into low-pressure, low-temperature zone
On-line Fuelling	No	Yes

OTHER POWER REACTOR TYPES

Although less common, some other reactor types currently in operation are described below.

Gas-Cooled Reactors use graphite as moderator, either natural or enriched uranium fuel, and carbon dioxide or helium gas as coolant. Most of the early reactors of this type were built in the United Kingdom and are called Magnox reactors (Magnox refers to the magnesium oxide alloy used to sheath the uranium fuel). Britain has developed an advanced gas-cooled reactor. Although carbon dioxide was used in some early commercial reactors, helium has become the gas of choice because it is chemically inert. The helium is pressurized to about 6.9 Megapascals and its temperature is raised to about 700 °C, which is considerably higher than for light and heavy-water reactors. A promising aspect for gas-cooled reactors is the potential for using the hot gas to directly drive a gas turbine, avoiding the use of steam. Negative features are that the graphite moderator is combustible, and gas can readily escape if a pipe bursts, increasing the probability of a loss-of-coolant accident.

Light-Water Graphite Reactors use graphite as moderator, enriched uranium fuel and boiling light water as coolant with a pressure tube design. These are used in Russia including the now infamous RBMK reactors at Chernobyl.

INTERNATIONAL SUMMARY

In 1987 there were 418 power reactors in 26 countries with a total electrical generating capacity of approximately 308,000 MW. By 2000, nuclear power had grown to 433 power reactors in 31 countries with an electrical generating capacity of approximately 349,000 MW representing a growth of about 13% over 13 years (see Table 7-2). It is clear that nuclear energy is a key component of today's global energy supply, and is growing relatively rapidly, particularly in developing countries.

Table 7-2	THE WORLD'S NUCLEAR CAPACITY		
Year		1987	2000
No. Countries		26	31
No. Reactors		418	433
Total Capacity (MW)		308,000	349,000

Table 7-3 shows the number of nuclear reactors and their capacity for the seven countries with the largest nuclear power programs. The USA has the world's largest nuclear program with over 100 reactors with a cumulative electrical capacity of 97,000 MW. Canada is in ninth position. It has fallen from sixth position because of the shutdown of seven Ontario reactors for extended maintenance.

France is far ahead of all other large countries in terms of the percentage of total electricity generated by nuclear (75%). The environment has been a big winner, as France reduced annual carbon dioxide emissions to 1.7 tonnes per capita, compared to 3 tonnes for

Table 7-3	NUCLEAR POWER BY NATION (2000)			
Rank	Country	No. Reactors	Capacity (MW)	%Total Electricity
1	USA	104	97,000	20
2	France	59	63,000	75
3	Japan	53	44,000	35
4	Germany	19	21,000	31
5	Russia	29	20,000	14
6	Korea (S.)	16	13,000	43
7	UK	35	13,000	25
8	Ukraine	14	12,000	44
9	Canada	14	10,000	12

FRANCE – THE NUCLEAR SHEIK OF EUROPE

France, in contrast to her neighbouring countries, has adopted a very proactive nuclear energy policy. In 1973, the Arab oil embargo and the huge increase in oil costs left France, which has only limited oil reserves, in a vulnerable position. To avoid a repetition of such a humiliating situation, France embarked on a major program to become energy self-sufficient. With ready access to uranium supplies, France decided to make her electrical supply primarily nuclear. Since 1973, France has constructed 59 nuclear plants, and now not only supplies 75% of its own electricity but also exports electricity to England, Germany, Switzerland, and Italy. It is interesting to note that Switzerland, which has placed a moratorium on further nuclear power plant construction, has no qualms about importing nuclear electricity. The cost of electricity in France is now one of the lowest in Europe. For the decade up to 1993, average electricity prices declined 20 to 30 per cent for industrial users and about 20 per cent for residential customers. France is now the largest electricity exporter in the world and has become the nuclear sheik of Europe.

Germany and five tonnes for the USA. Emissions of sulphur and nitrous oxides, leading contributors to acid rain, smog, and respiratory diseases, have been reduced by a factor of more than five.

Table 7-4 summarizes the different reactor types by number of plants and total energy capacity. It is seen that pressurized light-water reactors are the dominant reactor type in the world, comprising about 57% of all reactors and 60% of total generating capacity. Heavy-water reactors such as the CANDU are in third place, comprising 8.5% of reactors and 5% of nuclear electrical capacity.

The use of nuclear energy internationally continues to grow and at the beginning of 2001 there were 31 reactors under construction, most of the in the rapidly expanding economies of the Pacific rim, such as Korea, China, Japan, and Taiwan. Although there have been no new reactor orders in the USA in the past decade, the use of nuclear has been undergoing a small renaissance. The efficiency of operation and capacity factors have improved markedly so that the amount of nuclear electrical generation in 2000 was 20% higher than in 1990 — the equivalent of having built 19 new nuclear plants. In 1999, for the first time in well over a decade, the production costs of nuclear were lower than for coal (as well as oil and natural gas). In

Table 7-4	REACTORS SUMMARIZED BY TYPE (2000)		
	Reactor Type	# Units	Net MW
	Pressurized (Light) Water Reactors	250	222,000
	Boiling (Light) Water Reactors	93	80,000
	Heavy Water Reactors	37	20,000
	Graphite-Moderated Reactors	15	14,000
	Gas-cooled Reactors	35	12,000
	Other	3	1,000
	Total	433	349,000

addition, many of the older plants are planning major upgrades so they can obtain life extension permits and operate for many more years, rather than closing.

REACTORS THAT DO NOT GENERATE ELECTRICITY

Although the main use of nuclear reactors is to generate electricity, they have other applications as well. There are approximately 280 research and isotope production reactors operating in 54 countries. Some of the research applications of these reactors, which are much smaller than those that generate electricity, are described in Chapter 16.

Nuclear power is particularly well suited for craft that must travel for long periods without refuelling. Currently there are about 250 ships, mostly submarines with some icebreakers and aircraft carriers, powered by more than 400 small nuclear reactors. These ships all use pressurized-water reactors with special fuel to enable them to go

STEAM GENERATOR EXPORTS

A steam generator transfers heat from the reactor coolant to a separate circuit where it turns water into steam, which is then used to drive turbines to produce electricity. Like all types of boilers, they tend to corrode over many years of service and must be replaced periodically. Babcock and Wilcox Canada of Cambridge, Ontario, has become a leading manufacturer and exporter of steam generators. It supplies them not only for CANDU reactors around the world but also as replacement steam generators for US PWR reactors.

about ten years between refuelling. Due to the end of the cold war, arms reduction, and obsolescence, many nuclear submarines are being decommissioned.

Another type of craft that requires energy for long voyages is the space probe. On October 15, 1997, at Cape Canaveral, Florida, the dark night sky was suddenly lit by the roaring flames of a rocket lifting off the launching pad. The Cassini space probe was leaving on a 3.5 billion kilometre voyage to explore Saturn and its rings and moons that will last eleven years. Vital to the success of the mission are its nuclear thermoelectric generators that convert decay heat from the radioisotope plutonium-238 into electricity to power the space-craft and its 12 scientific instruments.

The US National Aeronautic and Space Administration has launched 23 deep-space probes during the past 30 years that used plutonium-238 for power. Plutonium-238, unlike its cousin isotope plutonium-239, is not fissile. Radionuclide energy is necessary for such long voyages because at those distances the sun is too weak to provide energy. Nuclear power is the only form of energy supply that can be used for long missions or for extended periods such as on space stations.

In 1978, Canada had to clean up the radioactive debris from Russia's Cosmos 954, a nuclear-powered satellite that fell down over the Northwest Territories. Canada has never launched any nuclear ships or satellites.

Apart from marine propulsion, research, and isotope production, nuclear reactors are seldom used for non-electric applications. There are some energy-intensive applications, however, to which they are well suited which we may see in future years including desalination of sea water in arid areas, extracting and refining oil from the subsur-face and from tar sands, and district heating. There is already some experience with cogeneration, that is, putting the residual heat from electricity generation to beneficial use. This was being done to a small degree at the Pickering station where low-grade heat was used in a fish farm, and at Bruce where an energy park has been established.

Chapter Eight

Safety:
The Prime Imperative

The accidents at Three Mile Island and Chernobyl have made everyone aware of the serious ramifications of a nuclear accident. Nuclear reactors pose a unique hazard that is not associated with fossil-fuel power plants, namely, they have the potential for rapid overheating which can damage the reactor and release dangerous radioactive materials.

In this chapter we consider the basics of reactor accidents and the physical and institutional mechanisms in place to protect the public and environment.

REACTOR SAFETY

Reactor safety is dominated by the three Cs: control, cool, and contain. The first imperative is to maintain control of the nuclear chain reaction at all times and be able to immediately stop it in an emergency. The second is to always cool the fuel to prevent it from being damaged. The third is to contain any radioactivity that might be released in an accident inside the reactor building so that it cannot harm the public or the environment.

The second of these principles is complicated by the fact that the fuel continues to emit heat even after the chain reaction is halted due to the radioactive decay of the fission products in the fuel. This heat decays away fairly rapidly with time but, in the absence of any heat removal, it is sufficient to melt the fuel for the first few hours after shutdown. Thus, the fuel must be cooled even after the reactor is shut down.

Let us look at how a reactor accident might happen. Rapid over-heating of the fuel could occur if there were, for example, a large pipe rupture or pump failure so the coolant leaks from the system or stops circulating. Heat would rapidly accumulate that could melt the core and also create large quantities of steam that could carry radioactivity to the environment.

The CANDU reactor has been designed with a number of barriers to prevent the escape of radioactivity in the event of an accident. First, the fuel is in the form of ceramic pellets with a melting point of 2840 °C. The pellets, in turn, are encased in Zircaloy metal. The fuel bundles lie inside a closed heat-transport system, and the reactor is surrounded by a massive concrete containment building. Finally the reactor is surrounded by a 1-kilometre exclusion zone inside which no homes are permitted. This multi-barrier 'defence-in-depth' approach is a fundamental tenet of CANDU reactor safety design (Figure 8-1).

It must be stressed that even in the most spectacular and far-fetched type of accident it is impossible for a reactor to explode like a nuclear bomb. The low concentration of fissile uranium-235 used in reactors and a number of other factors prohibit this from being scientifically possible.

Although the CANDU safety approach emphasizes accident prevention, it also recognizes that failures may occur and that systems and humans are not perfect. The CANDU reactor is designed to minimize the impact of such failures and imperfections. The Canadian approach emphasizes the separation and independence of three main systems: protective devices, operating systems, and containment.

As discussed in Chapter 6, the reactor's regulating system uses several different devices to control the nuclear chain reaction. In addition, there are totally independent and separate systems for the sole purpose of handling accidents. They are actuated only if the regular systems are unable to handle the safe cooling and/or shutdown of the

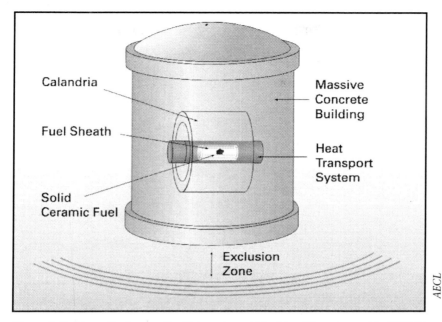

AECL

Figure 8-1: Schematic showing the 'defence-in-depth' safety philosophy used in CANDU reactors. Radioactive materials are contained in ceramic pellets. The pellets are contained in the Zircaloy tubes (fuel sheaths) of the bundles. The bundles are contained in the pressure tubes of the heat transport (cooling) system. The pressure tubes pass through the cala ndria which is in a massive concrete containment building. The reactor is surrounded by an exclusion zone.

reactor. CANDU reactors are among the few international reactors that have two independent fast shutdown systems that are physically separate and have their own sensors and power supplies.

Shutdown systems: Each CANDU reactor has at least two of the following three independent systems that can shut down the nuclear reaction in an emergency. First, the moderator can be dumped out of the core into a tank below the reactor, thus stopping the nuclear reaction. This method was used in the first CANDUs, Pickering A and earlier models, however, it was realized that the moderator was a good heat sink for excess heat in accident conditions, and this method is no longer used. Second, cadmium shutoff rods can be inserted vertically into the reactor to absorb neutrons and stop the nuclear reaction. They operate under the action of gravity, assisted by springs; power is not needed for them to function. The third system involves the injection of gadolinium nitrate dissolved in heavy water into the moderator through horizontal perforated pipes. The gadolinium absorbs neutrons and 'poisons' the reaction. To ensure redundancy, each of these systems is completely independent of the others and has their own power supply. Furthermore, they are physically different types of systems (the rods are solid and operate vertically; the "poison" system is liquid and operates via horizontal pipes) and they are even serviced by different maintenance crews.

Emergency Core Cooling System (ECCS): Should the normal systems that provide core cooling fail, reactors have an emergency core cooling system. Water stored in a reservoir would be injected under pressure over the fuel to remove heat.

Containment has the dual function of containing the release of radioactivity and also suppressing the pressure surge of any steam release. First, the reactor buildings themselves are constructed of massive steel-reinforced concrete up to 1.8 metres thick as shown in Figure 8-2. In a single-unit reactor (the CANDU 6) the containment building water-dousing systems and air coolers provide short and long-term pressure suppression. In multi-unit systems, containment is provided by each reactor building. In addition, each reactor is connected to a vacuum building (see Figure 2-5) maintained at a negative pressure. Should there be a rise in pressure from a break in the heat-transport system in any reactor building, valves open and vent the steam and pressure to the vacuum building, where a dousing system condenses and cools the steam.

AECL

Figure 8-2: The 1.8-metre-thick containment wall of the Unit 1 CANDU reactor in Qinshan, China. The photo was taken just before the concrete was poured. Note the large number of steel reinforcing rods used to make a very strong structure.

REACTOR ACCIDENTS

Since 1952 there have been serious accidents in about ten of the approximately one thousand power and research reactors operated throughout the world. Table 8-1 summarizes these accidents.

The first major nuclear accident in the world occurred at Chalk River in December 1952 with the NRX reactor. Due to operator errors, the reactor shutdown rods remained stuck out of the core during a

Table 8-1
REACTOR ACCIDENTS

Reactor	Type	Date	Deaths	Public Exposure
NRX (Canada)	Research	1952	None	Minor
Fermi-1 (USA)	Prototype	1966	None	None
Lucens (Switzerland)	Research	1969	None	Very small
Browns Ferry (USA)	Power	1975	None	None
Three Mile Island-2 (USA)	Power	1979	None	Limited
St. Laurent-2 (France)	Power	1980	None	Minor
Chernobyl-4 (USSR)	Power	1986	31	Major
Vandellos-1 (Spain)	Power	1989	None	Very small
Monju (Japan)	Prototype	1995	None	None

power surge so that the fuel overheated, melted, and released significant radioactivity into the reactor building, as well as causing major damage to the reactor core. The nuclear reaction was stopped by opening a valve and removing some of the moderator. Extensive cleanup was required. No one was injured in the accident and only minor radioactivity was released to the environment, but a culture of safety was thrust onto the Canadian program in its early stages. Fourteen months afterwards, the reactor was recommissioned and operated until 1992.

The Three Mile Island accident was caused by loss of coolant and was further compounded by subsequent operator error. It was initiated by a faulty valve in the cooling system that stayed open when it should have closed. The operators were unaware that this particular valve was open because poorly designed alarm systems in the control room did not give them this information. They then made the significant mistake of manually turning off the Emergency Core Cooling System. As a consequence, fuel in the core was not cooled and a

PRESIDENT JIMMY CARTER

At the time of its accident (1952), NRX was the world's largest research reactor and was being used to test reactor fuel for the US Navy program. For this reason, the cleanup involved both Canadian and US servicemen, one of whom was Jimmy Carter, future President of the USA. Carter was educated as a nuclear engineer and at the time was an officer in the fledgling nuclear arm of the United States Navy. As President in 1979, he also dealt with the Three Mile Island crisis. Thus, he had close involvement with the worst nuclear reactor accidents in both Canada and the USA.

substantial portion of it melted. They only realized their error when radiation alarms in the control room sounded. They turned the Emergency Core Cooling System back on but by then it was too late and there was substantial fuel melting and core damage.

The radiation was almost all confined to the reactor containment. The public in the immediate area was exposed to only a very small radiation dose, equivalent to a small fraction of the yearly natural background dose. Although the reactor suffered significant damage, none of the reactor operators and no member of the public were injured. Nevertheless, a media storm erupted causing considerable psychological trauma to the public in the area. The lack of credible information from the reactor owners and the vacillations of the state government concerning evacuation created a state of fear and anxiety that was totally unwarranted. President Carter visited the site; he appointed competent individuals from his government to manage the situation, and generally helped restored public confidence.

The worst nuclear accident in history occurred at Chernobyl, Ukraine, on April 26, 1986. The operators of a Russian RBMK reactor were conducting an experiment to test a safety system that involved running the reactor at very low power, where it was unstable and difficult to control. In the space of a few seconds they lost control and a sudden huge increase in power caused a steam explosion that destroyed the reactor. The moderator, which was made of graphite, a form of carbon, caught fire. Chernobyl had inadequate containment for this size of accident so large quantities of radioactive materials were released. The moderator continued to burn for weeks, sending a plume of radioactive smoke into the atmosphere and creating serious problems for firefighters, several of whom received fatal radiation doses.

IAEA / Vadim Mouchkin

Figure 8-3: The Chernobyl reactor number 4 (middle with columns) contained in its "sarcophagus" to prevent release of radiation. Reactor number 3, on the left, operated until the end of 2000.

The accident took place during the Soviet Union era and it was one of the factors that eventually led to the demise of that regime. It was kept secret for some days and only discovered in the Western world during routine radiation monitoring at a Swedish nuclear plant. People in the immediate area of Chernobyl were not informed and were not evacuated for some days. The Soviet authorities, in contrast to the people at the scene dealing with the accident, were confused, ineffectual and generally incompetent in their management of the accident.

Air currents carried the cloud from Chernobyl both east and west, reaching Canada's east coast on May 6 and the west coast on May 7. The highest radiation concentrations in Canada occurred in May and by the end of June the levels had returned to normal. Health Canada estimated that the Chernobyl accident led to the typical Canadian receiving a radiation dose of 0.00028 mSv, a negligible dose compared to typical variations in natural background (average annual background exposure in Canada is about 2.6 mSv. The potential long-term radiological impact of this accident on the local population is discussed in Chapter 4.

It is worth noting that there are at least two major (and many minor) differences between the CANDU and Chernobyl reactors. First, a CANDU reactor does not contain a combustible moderator. The moderator and coolant consist of water and the CANDU reactor structure is made mostly of concrete. In contrast, the Chernobyl reactor used a graphite moderator which is composed of carbon. A CANDU reactor cannot catch fire and burn as an RBMK reactor can.

The second difference is the quality of the containment. Studies have shown that if the Chernobyl reactor had been equipped with a thick concrete containment such as is standard in CANDU stations (see Figure 8-2), very little radioactivity would have escaped. This does not even take into account the additional protection offered by the vacuum building.

Each of the above two factors alone would have prevented significant radiation releases in the Chernobyl accident. In addition, the CANDU has other differences that minimize the potential for and impact of a Chernobyl-type accident, including independent shutdown systems and operator training.

Reactor accidents such as at AECL's NRX reactor, Three Mile Island, and Chernobyl have shown that serious events almost always have a substantial component of human error. Thus, a significant CANDU safety feature includes the training of reactor operators to rigorous standards set by the regulatory agency, the Canadian Nuclear Safety Commission (CNSC), which includes the use of sophisticated simulators. The CNSC maintains permanent staff at each of the nuclear stations. In addition, elaborate emergency response plans have been developed that coordinate the activities of municipal, provincial, and federal agencies.

Canada has been an international leader in developing a systematic approach to minimizing risk in nuclear reactors. This concern with safety was a legacy of the 1952 NRX reactor accident at Chalk River, which formed a wake-up call early in the history of the CANDU. No nuclear plant worker in Canada has lost time from the job as a result of exposure to radiation in over 35 years, and no member of the public has suffered injury or death due to a reactor accident in Canada

INTERNATIONAL NUCLEAR REGULATION

Canadian nuclear regulations benefit from the exchange of information and ideas with other countries that have nuclear programs. In

IAEA

Figure 8-4: The International Atomic Energy Agency headquarters in Vienna, Austria. This United Nations organization is responsible for ensuring the safe and peaceful uses of nuclear technology.

particular, Canada participates in a number of international organizations that study the biological and environmental effects of radiation.

The International Commission on Radiological Protection (ICRP) was formed in 1928 to provide basic recommendations and guidance on radiation protection. Headquartered in London, England, it dealt initially with exposure to medical X-rays and radium. The commission was reorganized and given its present name in 1950; it consists of 13 scientists from the international community. In addition, there are four permanent committees of scientists with expertise in different relevant specialties. Working groups are set up from time to time to deal with special issues and call upon experts from around the world, as necessary. ICRP members are chosen on the basis of their individual merit in medical radiology, health physics, genetics, and other related fields, with regard to an appropriate balance of expertise rather than to nationality.

The ICRP issues reports with recommendations on various aspects of the protection of humans against all sources of ionizing radiation. The ICRP is a very influential organization and its recommendations regarding permissible dose limits form the basis of regulations in most countries.

In January 1946, the General Assembly of the United Nations created an Atomic Energy Commission with the purpose of promoting the peaceful use of nuclear energy, setting standards for radiation safety, and developing safeguards against the proliferation of nuclear weapons. The Commission was renamed the International Atomic Energy Agency (**IAEA**) in 1957.

A major function of the IAEA is to promote the exchange of information on nuclear safety between its members. One important example of its work is the establishment of a system through which a country can rapidly inform others when it has a nuclear accident. As we have seen, this did not happen in the case of Chernobyl.

The International Nuclear Events Scale shown in Table 8-2 is an integral part of this system. It also is of assistance to journalists and the public in understanding the significance of reactor accidents. All the IAEA countries including Canada have agreed to use it in reporting accidents to each other. As many minor incidents are reported in the media in a sensationalistic and frightening manner, the use of the International Events Scale would help place such incidents in perspective.

The United Nations Scientific Committee on the Effects of Atomic Radiation (UNSCEAR) was established in 1955 because of concern about radioactive fallout from atomic-bomb testing in the atmosphere. The committee, which consists of 70 to 100 scientists (physicists, biologists, geneticists, medical doctors, and others) from more than 20 countries including Canada, meets once a year. About every five years, UNSCEAR issues a document reviewing and updating scientific information on the levels of radiation to which humans are exposed and the biological effects of this exposure. In 2000, UNSCEAR issued a comprehensive review of the effects of radiation on health all over the world, including the area around Chernobyl.

The World Association of Nuclear Operators, representing nuclear utilities from 33 countries with responsibility for operating over 400 reactors, was established in 1989 in response to the Chernobyl accident in 1986. Operating as a voluntary organization, rather than a regulatory authority, its objective is to improve nuclear safety through international cooperation and sharing of information amongst its members. It has four regional centres located at Atlanta, Moscow, Paris, and Tokyo.

Each industrialized country also has its own national committee on radiation protection. In the USA, the National Academy of

Table 8-2
THE INTERNATIONAL NUCLEAR EVENT SCALE

Level	Public Exposure	Examples
7-Major Accident	Major	Chernobyl
6-Serious Accident	Significant	
5-Accident - Off-site Risk	Limited	Three Mile Island
4-Accident Mainly in Reactor	Minor	St. Laurent, NRX
3-Serious Incident	Very Small	Vandellos, Lucens
2-Incident	None	Monju, Browns Ferry
1-Anomaly	None	

Sciences has established committees on the Biological Effects of Ionizing Radiation that are funded by the Environmental Protection Agency. The committees are staffed by scientists from universities, hospitals, and national laboratories. In Canada this role is filled by the Advisory Committee on Radiological Protection, which reports to the Canadian Nuclear Safety Commission.

Safety in the design and operation of CANDU reactors has continually evolved and improved over the more than 35 years since the first prototype was constructed. Design and operation are constantly reviewed and major research programs are undertaken by AECL, Ontario Power Generation, and other organizations. The Canadian program also benefits from the operating experience of nuclear reactors around the world through the Institute of Nuclear Power Operations in Atlanta and the International Atomic Energy Agency in Vienna. As well, the Canadian nuclear utilities and three international utilities with CANDU nuclear reactors (Argentina, Korea, Romania) share information and conduct joint research through the CANDU Owners Group.

CANADIAN NUCLEAR REGULATION

Laws governing the use of radioactive materials have been promulgated at both the federal and provincial levels in Canada. The Atomic Energy Control Act, passed in 1946 and administered by the Atomic Energy Control Board, was the original federal law governing radiation matters. In 2000, a new act was implemented, the Nuclear Safety and Control Act, establishing a new nuclear regulatory agency, the Canadian Nuclear Safety Commission (CNSC). The CNSC regulates the use of nuclear energy and materials to protect health, safety, security, and the environment, and to respect Canada's international commitments on the peaceful use of nuclear energy.

Agnes Bishop (1938-): Following a distinguished career as a physician specializing in pediatric medicine, Dr. Bishop became the first woman president of the Atomic Energy Control Board in 1994. In this role she guided the introduction of the new Act governing nuclear safety and managed the transition of the Atomic Energy Control Board to the Canadian Nuclear Safety Commission, retiring as its first President in 2000.

CNSC

Responsibility

Today, the CNSC's responsibilities are as wide-ranging as the use of radioactivity. Through a comprehensive licensing process, it regulates all aspects of Canada's nuclear industry. There are two main areas: the nuclear fuel cycle and the use of nuclear substances. These include: power reactors, research reactors, particle accelerators, uranium mine/mill facilities, uranium refining and fuel fabrication facilities, heavy-water plants, radioactive waste management facilities, and the use and production of nuclear substances and radioactive gauges.

The Commission's duty is to ensure that all nuclear substances in Canada are used so there is minimal adverse health or environmental impact. It is also responsible for nuclear security aspects, transportation, and export of nuclear substances. In addition, the CNSC has some responsibilities regarding insurance against nuclear liability. It contracts for research in areas relevant to its regulations, and in some special circumstances it supervises decontamination projects.

Regulatory Philosophy

The underlying principle governing the CNSC's approach to regulation is to set technically and publicly acceptable standards with respect to public and worker risk to radiation exposure. It is then up to the nuclear facility owner (i.e., the licensee) to determine how these standards will be met. The role of the Commission is to ensure that the licensee lives up to its responsibilities.

Canada and, for example, Great Britain share this philosophy of giving the licensee relative freedom in choosing how to meet the regulations. The US regulator, in contrast, is much more prescriptive and imposes highly detailed regulations. The Canadian approach is called performance-based licensing, whereas the US approach is referred to as prescriptive licensing. CISC's approach is felt to allow for greater flexibility in the engineering and design of nuclear facilities, and encourages innovation and improvements, rather than restricting them through specific, prescriptive regulations.

In addition, the CNSC regulates the nuclear industry according to the As Low As Reasonably Achievable (**ALARA**) principle. That is, actual exposures must not only meet the prescribed limits, but should be kept as far below them as possible, economic and social considerations being taken into account.

As in most countries, Canada's regulations are based on the recommendations of the International Commission on Radiological Protection.

CNSC Structure

The Commission consists of up to seven people appointed by the federal government. The Chair of the Commission is also the chief executive officer of the CNSC. A staff of about 450 are organized in various directorates spanning the Commission's areas of responsibility.

CNSC's headquarters are in Ottawa and field offices are located throughout Canada and at reactor sites. In addition, the CNSC operates a laboratory in Ottawa that performs radiochemical analyses necessary for their independent evaluations of licensees, as well as maintenance, testing, and calibration of the instruments used by field staff.

The CNSC reports to Parliament through the Minister of Natural Resources Canada. As this is the same Minister to whom AECL reports, some critics feel that there is a conflict of interest, and occasionally recommendations are made to have the CNSC report through an independent Minister, such as the Minister of the Environment.

The Advisory Committees on Radiological Protection and Nuclear Safety provide expert opinion and independent guidance to the Commission in formulating standards and regulations. The committees are composed of eminent scientists and engineers from outside, with support from CNSC staff. In setting radiation limits, the CNSC is also influenced by the international bodies discussed in the previous section, particularly the International Commission on Radiological Protection, and by other federal departments such as Health Canada.

Licensing Process

A licence must be obtained from the CNSC by anyone who wishes to possess, use, import, or export nuclear materials (uranium, thorium, heavy water, and radionuclides) or by anyone wishing to operate a nuclear facility. Examples of a nuclear facility include a uranium mine and mill, refinery, conversion plant, fuel fabrication plant, heavy-water plant, nuclear reactor, or waste management facility.

The licensing process for all major nuclear facilities such as reactors or waste disposal facilities includes, but is not limited to, three major stages:

1. Site Licence, which requires provincial and federal environmental assessments and public hearings.

2. Construction Licence, which requires a preliminary safety report

3. Operating Licence, which requires a final safety report

In addition, there are ongoing inspections and periodic licence renewals. For example, there are CNSC staff assigned full time at the Pickering, Bruce, and Darlington nuclear generating stations. They monitor compliance with licensing requirements and perform routine inspections. They have the authority to order any reactor to reduce power or shut down if licencing conditions are not being met.

Licensees are required to keep detailed records of their operations including radiation exposures of all their employees and concentrations of all radioactive emissions. Worker exposures, measured by devices called dosimeters, must be reported to the National Dose Registry operated by Health Canada. By 1998, the registry contained the cumulative exposure records of over 500,000 individuals who worked in more than 80 different occupations.

At the end of the useful life of a nuclear facility, it must be **decommissioned** in a manner that is acceptable to the CNSC. This generally involves the restoration and clean-up of the site for unrestricted use, or managing it until it no longer poses a hazard to human health or the environment.

Nuclear Substances and Radioactive Materials
The CNSC issues a large number of licences for use in medical diagnosis and treatment as well as in industry and research. In 1988, they issued 4860 licences and performed about 2800 inspections to ensure that the licence conditions were being met. The number of licensees

has decreased in recent years. In 1998, there were about 3700 licences which were distributed across Canada as shown below.

Table 8-3	RADIONUCLIDE LICENSES (1998)	
	Type of User	No. of Licenses
	Hospitals and medical	801
	Universities and educational	287
	Governments	370
	Commercial*	2,242
	Total	3,700

*includes oil well logging, radiography, gauging, static eliminators, etc.

Transportation

About one million packages of radioactive material are shipped in Canada each year. The CNSC regulates the packaging, preparation for shipment, and receipt of these radioactive materials through the Packaging and Transportation of Nuclear Substances Regulations. In addition, the CNSC cooperates with Transport Canada in regulating the shipments of radioactive materials under the Transportation of Dangerous Goods Act.

Shipments of radioactive materials can be made only in approved packaging or by special arrangement on a case-by-case basis. To receive approval, packaging must meet established performance standards, which for shipment of highly radioactive materials are very stringent. This type of packaging must withstand a series of tests including heavy impact, fire, and submersion in water. Some packages that have been certified in this way have survived the test of being struck by a locomotive traveling at 165 kilometres per hour.

Accidents in which significant amounts of radioactive material are spilled are very rare. To date, no personal injuries due to radiation exposure in shipments of radioactive materials have occurred in Canada.

Nuclear Liability Insurance

The CNSC is also responsible for administering the Nuclear Liability Act of 1970, which established the amount of basic public liability insurance to be maintained by the operator of each nuclear facility. Currently the operators of nuclear power plants must carry $75 million commercial third-party liability insurance. This coverage is solely to meet claims by the public in the event of a nuclear accident;

it cannot be used to repair the damaged reactor facilities. The Nuclear Insurance Association of Canada, a consortium of insurance companies licensed to do business in Canada, is the only approved source from which reactor owners can obtain liability insurance. The cost to the reactor owner for annual premiums depends on factors such as population density and land values near the reactor.

The Act also provides the federal government the power to establish a commission to deal with situations where the claims could exceed that amount in the case of a nuclear accident. After three decades of reactor operation in Canada, no insurance claims have yet been made. Recently there has been discussion in the government about raising the $75 million to a much higher level and this will probably occur in the near future.

Safeguards

Safeguards can be defined as a system of treaties, international agreements, physical measurements and on site inspections of nuclear facilities aimed at preventing the misuse of nuclear technology and the diversion of nuclear materials for weapons purposes. The overriding policy of the Government of Canada is that nuclear materials, equipment, and technology be used for peaceful purposes only. This policy has evolved in two main stages. In 1974, India exploded a nuclear bomb that used plutonium produced in research facilities supplied by Britain, Canada, and the US. Canada immediately cut off all nuclear technology exchange with India. Shortly thereafter, Canada began to insist that customer countries provide binding assurances that Canadian nuclear materials, equipment, and technology will not be used to produce nuclear weapons. In 1976, Canada further required that all customer countries that are non-weapons states must ratify the *Non-Proliferation Treaty* or otherwise accept international safeguards of their entire nuclear program, both present and future. To date, 121 nations have signed the Non-Proliferation Treaty.

The CNSC is responsible for ensuring that Canada adheres to international protocols on nuclear safeguards. These programs help ensure that nuclear fuel is not stolen or otherwise diverted from its intended peaceful uses. The objective is to prevent the manufacture of nuclear weapons. Canada is a signatory of the Non-Proliferation Treaty. The CNSC, jointly with AECL, administers the Canadian Safeguards Support Program, which assists the IAEA in improving methods of ensuring safeguards, including improving safeguard equipment installed at the CANDU 6 reactors.

Other Canadian Jurisdictions

Nuclear regulation covers a broad spectrum of activities, many of which relate to areas under provincial jurisdiction, or are within the areas of expertise of other federal departments. To ensure good communication and cooperation with these agencies, the CNSC has developed a joint regulatory process. At the provincial level, for example, the CNSC coordinates with the various ministries responsible for health, labour, environment, and natural resources.

Exposure Limits

The CNSC sets the exposure limits for what are called Nuclear Energy Workers, which includes employees at nuclear reactors. Whole body radiation exposure is not to exceed an annual limit of 20 mSv, averaged over five years, with the maximum dose in any given year not to exceed 50 mSv. This limit is based on the recommendations of the International Commission on Radiological Protection and on having radiation risks comparable to the risks workers face in other safe industries.

The exposure limit for members of the public is established at one mSv per year, which is based on the level of risk people are regularly exposed to in their everyday lives from natural background radiation. This limit was recently decreased from five mSv per year in conformance with the recommendations of the International Commission on Radiological Protection. As described in Chapter 4, there is considerable debate about the effects of low-levels of radiation. Thus, lowering the exposure limit added a degree of safety, that is, conservatism. Canadian nuclear reactors and research facilities readily meet this new limit.

NUCLEAR CRITICS

In Canada, we recognize that an effective opposition is essential for the proper functioning of parliamentary democracy. The opposition, by questioning, analysing, and criticizing the government's decisions and policies, ensures that complacency and the arbitrary use of power are avoided. Cases of government abuse, incompetence, and waste are brought before the public and media for the judgement of the voters. Similarly, one can argue that nuclear critics are essential to the proper functioning of our nuclear safety system. The fact that critics are always looking over the shoulders of the nuclear industry and its regulators keeps them from lapsing into complacency.

Norman Rubin (1945-): one of Canada's foremost nuclear critics, was born and educated in the US in science and music. He came to the University of Toronto in 1974 as a lecturer and later joined Energy Probe as a researcher. He has written extensively on nuclear issues and has testified at numerous public inquiries and commissions. In addition, he has appeared on most of Canada's radio and television programs that deal with news and current events.

Energy Probe

Canada is fortunate to have a small number of dedicated nuclear critics, some of whom have devoted a large portion of their working lives to the often thankless task of critiquing nuclear matters. Over the years there have been a great many commissions and committees that have studied various aspects of the nuclear industry. Each of these studies has included interventions by nuclear critics as part of an open and transparent process. Nevertheless, most of these studies have resulted in decisions that are contrary to the desires of the critics in terms of allowing existing nuclear facilities to continue operation or for new ones to be constructed.

In 1987, Energy Probe, the City of Toronto, and others launched a legal challenge to the Nuclear Liability Act claiming that the Act was unconstitutional. In a judgement released on March 23, 1994, the court dismissed the actions and awarded legal costs to the defendant. In his summation the Judge stated that the plaintiffs failed to show that the Nuclear Liability Act had caused less safety in the operation of nuclear reactors or resulted in increased risk to the public.

The critics have enjoyed considerable success in establishing a strong presence with the media. Every nuclear-related event receives close scrutiny by the press and by the critics, whose comments are extensively reported. While the critics sometimes make technically inaccurate statements and often exaggerate for effect, what they say is not as important as the fact that they are there.

Chapter Nine

Nuclear Power and the Environment

With our major cities sitting under pallid umbrellas of smog, the global temperature inching remorselessly higher, and natural spaces falling under the bulldozer's blade, the issue of environment is on everyone's mind. In this chapter we look at some of the environmental impacts of nuclear power and compare them to those of other energy sources. In particular, the issues of air pollution, global warming, reactor emissions, and low level nuclear wastes are addressed, leaving the issue of high-level wastes for the following chapter.

ACID RAIN AND AIR POLLUTION

The burning of fossil fuels, especially coal, produces many harmful air emissions. Sulphur dioxide and nitrogen oxides, for example, enter the atmosphere and cause sulphuric and nitric acid, respectively. Falling as acid rain, these gases have caused extensive damage to lakes and forests in Canada and elsewhere in the world. Hundreds of lakes have been rendered so acidic that they no longer support fish life. Trees are suffering blight in many parts of the world and acid rain is suspected of contributing to the problem. Nitrogen oxides combine with hydrocarbons in the presence of sunlight to cause ground level ozone, one of the most harmful air pollutants. In Europe, about 60% of sulphur-dioxide emissions and 35% of nitrogen-oxide emissions come from electricity generation.

In addition to sulphur and nitrogen oxides, the combustion of coal releases carbon dioxide, hundreds of volatile organic compounds, carbon monoxide, furans and dioxins, as well as trace amounts of heavy metals and radioactive elements into the air. Some of the carbon remains unburned, and is released as solid particulates, which are generally less than one five-thousandth of a centimetre in diameter. These particulates can float in the air for days with organic compounds and trace elements clinging to them. When inhaled, these particulates contribute to acute and chronic respiratory illnesses, including cancer. It is estimated that in Ontario alone air pollution causes 1,900 deaths each year, and coal-burning electrical plants make a significant contribution to this problem.

Each year in the United States alone, air pollutants from burning coal kill thousands of people (estimates range from 5,000 to 200,000), cause at least 50,000 cases of respiratory diseases, and result in several billion dollars of property damage. In 1975, it was estimated that about 26 deaths in the United States were caused per gigawatt-year (1 gigawatt = 1,000 MW) of coal electricity.

Introduction of sophisticated pollution-control technologies have significantly reduced some emissions, but such devices are expensive and have increased the cost of coal-produced electricity. Many pollutants escape the pollution devices, for example, carbon dioxide, a major global warming gas, cannot be captured by current pollution-control technologies.

A generally unrecognized fact is that burning coal emits thousands of times more radioactive particles into the atmosphere than do nuclear power plants. This occurs due to the natural radioactivity contained in coal (and all rocks on earth). For example, some years ago the UK's Central Electricity Generating Board estimated that it released 300 kilograms of uranium into the environment every day from burning coal.

In 1998, Ontario substantially increased power generation from coal-fired stations to replace the electricity no longer available from the seven nuclear stations shut down for extended mainte-

Mike Graston, Windsor Star

Figure 9-1: This cartoon makes the point that there are skeptics about global warming. It also warns us not to make scientific decisions based solely on our personal perceptions.

nance. Ontario then became one of the leading sources of air pollution in North America due to the chemicals and smog from these plants. Health authorities in Toronto have estimated that several hundred additional deaths annually can be attributed to this degradation of air quality.

Emissions from electricity-generating plants that use fossil fuels are now recognized as one of the most serious environmental threats in human history.

GLOBAL WARMING

Over hundreds of millions of years nature has been locking up carbon in underground deposits of coal, oil and natural gas. Now the industrial revolution is undoing this work in a matter of a century or so — a mere heartbeat in geological time — releasing enormous quantities of carbon dioxide into the atmosphere. At the same time the worldwide loss of forests is reducing the amount of carbon dioxide that is converted to oxygen by photosynthesis. As carbon dioxide levels increase, the atmosphere captures more of the solar infrared radiation that would normally be reflected back into outer space, thus increasing the atmospheric temperature. This so-called "greenhouse effect" is essential to life on earth, and without the blanket of warm air it creates, life would be very difficult to sustain. However, concentrations of CO_2 in earth's atmosphere have risen from less than 280

121

parts per million before the industrial revolution to about 367 parts per million in 2000, with an attendant increase in the global temperature of about 0.5 degrees Centigrade.

There is little doubt that both atmospheric carbon dioxide and global temperatures have increased. Most scientists believe that global warming is caused by human activities and that burning fossil fuels is a major contributor. It should be noted, however, that there are some skeptics who attribute the observed warming to other effects over which humans have no control, for example, changes in the sun's energy output.

Although some may think that warming would be good in colder areas of the world, the global consequences could be far from pleasant. Melting of polar ice caps and rising sea levels would dramatically affect coastlines and for some countries like Holland could be devastating. Global warming would also cause extreme weather such as hurricanes and killer heat waves and lead to tropical diseases moving northward. Many areas, like Canada's prairies, could be transformed into arid deserts.

Globally, about 25% of the carbon dioxide contribution to the atmosphere is caused by electrical generation using fossil fuels. Canada releases the equivalent of 618 million tonnes of CO_2 in greenhouse gases into the atmosphere each year with electricity generation contributing 17% of this total.

Carbon, a major constituent of coal, oil, and natural gas, burns according to:

$$C + O_2 \rightarrow CO_2$$

Thus, carbon dioxide emission is a fundamental consequence of burning fossil fuels to produce electricity. In fact, for every kilogram of carbon burned 3.67 kilograms of CO_2 are produced. Burning natural gas is better than burning coal as it produces about half the CO_2 for the same amount of energy. Natural gas consists largely of methane, however, and methane is some twenty times more effective than CO_2 in trapping heat in the atmosphere. Thus, the various leaks of methane to the atmosphere arising from drilling, transporting, and distributing natural gas reduce its advantage in combustion. Oil combustion is somewhere in between coal and natural gas. A quantitative comparison is given in Table 9-1. Interestingly, firewood is

Table 9-1	EMISSION OF CARBON DIOXIDE BY FOSSIL FUELS	
Fuel	CO_2 per Unit Energy Produced*	
Soft Coal	95	
Firewood	94	
Hard Coal	90	
Crude Oil	72	
Natural Gas	51	
* kilogram of CO_2 per 1,000 MW		

about the worst possible fuel in terms of climate change and has the added negative impact that it involves cutting down trees which would otherwise soak up carbon dioxide.

The government of Canada has done an extensive study of what can be done domestically to reduce greenhouse gas emissions. Several possibilities are under consideration mainly based on energy efficiency and energy conservation. Most experts, however, feel that Canada has little chance of meeting the Kyoto targets. In fact, Canada's greenhouse gas emissions, rather than decreasing, have increased since 1990.

Nuclear stations do not degrade the environment with greenhouse gases or other smog-producing pollutants, although minor amounts of radioactivity are released. Solar energy and wind energy, in common with hydro power and nuclear energy, are the only electrical-generation technologies that do not emit greenhouse gases. Unlike nuclear energy, these other energy sources cannot provide the large amounts of electrical power required in the industrialised areas of Canada. After all practical energy conservation and energy-efficiency measures have been taken, nuclear power appears to be the only method for achieving significant reductions in greenhouse gases.

REACTOR EMISSIONS

Nuclear generating stations release small quantities of radioactivity into the atmosphere and adjoining water bodies. These emissions are monitored to ensure they are kept well below levels that could cause harm to humans and the environment. The magnitude of the doses that members of the public may receive due to these emissions are so low they can not be measured directly to see if they are in regulatory compliance. Instead, the Canadian Nuclear Safety Commission has established Derived Emission Limits (DELs) by considering the exposure pathways through the environment by which radioactive emissions could reach members of the public. If emissions are kept below the specified DELs, calculations show that the dose to the most exposed member of the public will be less than the regulatory limit, that is, the whole body dose will not exceed one mSv in any year.

Airborne emissions include tritium, argon, iodine-131, and particulates. Water emissions include tritium and other beta and gamma radioactive emitters. The nuclear utilities and other provincial and federal agencies monitor all the possible paths by which radioactivity can enter the environment and the food chain in the vicinity of the reactors, including air, precipitation, milk, drinking water, vegetation, algae, and fish. Because radiation is easy to detect with appropriate instruments (see Appendix A), these reactor emissions can be tracked accurately.

The CNSC released data for the period 1988-1997 showing that the emissions from all Canadian nuclear power reactors were well below the mandated DELs. In fact, no emissions exceeded 1% of those values, a target that the nuclear utilities have voluntarily set themselves.

During the more than 30 years of Ontario Power Generation's nuclear operations, which involve more than 300 reactor years, there has never been a known injury to the public involving radioactivity. Furthermore, there has never been a release of radioactivity from any Ontario Power Generation nuclear generating station that resulted in a measurable radiation exposure to any member of the public.

OVERVIEW OF NUCLEAR WASTES

In Canada, nuclear wastes are divided into three main categories: spent fuel (also known as high-level waste), low-level waste, and historic wastes.

Spent fuel consists of fuel bundles which have been used in a nuclear reactor until they no longer produce power effectively. Although these bundles appear the same physically as before going into the reactor, spent fuel is highly radioactive, generates heat, and must be handled by remote control.

There are many definitions of low-level radioactive waste (**LLRW**), however, the broadest and simplest is that LLRW consists of all those radioactive wastes that are not high-level waste. In Canada this comprises all the radioactive waste which is not spent reactor fuel. LLRW includes a broad spectrum of chemical and physical forms such as filters and resins from cleaning a reactor's coolant, worn-out reactor components, protective clothing, and much more. Although LLRW contains many different radionuclides, its radioactivity is relatively small, contributing less than 1 per cent of the radioactivity at nuclear reactor stations; the other 99% is in the spent fuel. LLRW also includes the radioactive wastes from nuclear medicine, research and industrial applications of radiation.

Historical wastes are a special category of LLRW for which the generator no longer is in business or cannot be held responsible. These consist primarily of wastes in and near the town of Port Hope, Ontario, that accumulated between 1933 and 1988 from the uranium refinery and conversion plant operated by the former Eldorado Nuclear Limited.

To provide perspective to the problem of nuclear wastes, it is helpful to view them in the broader context of the non-nuclear wastes generated by society.

Toxicity
Every home in Canada generates wastes, which are hauled weekly from our curbside. In addition to items such as chicken bones and empty tooth-paste containers, they also contain hazardous materials such as used motor oil, batteries, oven cleaners, paint thinners, pesticides, tires, wood preservatives, drain cleaners, drugs, and much more. Approximately 1.9 kg of hazardous waste is produced per person per year, much of which winds up in municipal landfills. In addition, industries produce large quantities of hazardous waste.

Hazardous waste includes, for example, chlorinated organic compounds such as vinyl chloride and trichloroethylene as well as heavy metals including copper and lead. Vinyl chloride can cause liver cancer and neurological disorders, and lead can damage the nervous and reproductive systems. These wastes can cause damage to biological systems that is more severe than the effects of nuclear wastes. In particular, there are more than 1,500 (non-radioactive) substances that are known to cause cancer.

The toxicity of nuclear waste has been compared to that of other substances by Cohen (1990), who calculated the amounts of reprocessed US nuclear waste that would need to be ingested by a

Table 9-2	LETHAL DOSES FOR NUCLEAR WASTE	
	Years After Burial	Amount Causing Death (grams)
	1	0.3
	100	2.8
	600	28.0
	20,000	454.0

human to be lethal. The results are shown in Table 9-2. For comparison, Table 9-3 shows the amounts of various chemical compounds that would need to be ingested to be lethal.

Comparing Tables 9-2 and 9-3, it is seen that the toxicity of nuclear waste is not significantly different from that of other compounds that are commonly used in industry. If the decrease in toxicity with time is taken into consideration, nuclear wastes are actually less toxic than many non-nuclear compounds.

Table 9-3	LETHAL DOSES OF CHEMICAL COMPOUNDS	
	Compound	Lethal dose (grams)
	selenium compounds	0.3
	potassium cyanide	0.6
	arsenic trioxide	2.8
	copper	20.0

Dealing with nuclear wastes should be easier than dealing with non-nuclear toxic substances since nuclear wastes are kept isolated and contained, and become less toxic with time. In contrast, many non-nuclear toxic compounds such as arsenic trioxide, a herbicide and insecticide, are scattered intentionally on the ground where food is growing, and are even sprayed directly on fruits and vegetables.

Quantities

Let us compare the quantities of nuclear and non-nuclear wastes. The capacity of the Keele Valley landfill in Toronto, Canada's largest municipal landfill, is 20 million tonnes. Studies have shown that between approximately 0.1 and 3% of a municipal landfill consists of hazardous wastes, which would yield approximately 0.02 to 0.6 million tonnes of hazardous waste in the Keele Valley landfill. This is similar to the capacity of the proposed high-level nuclear waste vault (about 0.25 million tonnes of used fuel). A single nuclear repository would contain all of Canada's nuclear waste, whereas Keele Valley is but one of thousands of municipal landfills in Canada.

The situation is the same for landfills containing hazardous waste (rather than municipal waste). For example, the Safety Kleen hazardous-waste landfill near Sarnia, Ontario, will contain about 7.5 million tonnes of hazardous waste when it is full. This is approximately 30 times more waste, by weight, than will be contained in the proposed spent-fuel repository. The Safety Kleen facility is one of three hazardous-waste disposal facilities in Canada.

It is clear that society produces much more non-nuclear waste than nuclear waste of similar toxicity.

Duration of Toxicity
Some components of nuclear waste last a long time, for example, plutonium-239 has a half-life of about 24,100 years. Many people are concerned that scientists and engineers cannot build a disposal facility that would retain its integrity for this length of time, which exceeds the period of recorded history. The disposal of non-nuclear wastes, however, involves toxic time spans of even longer duration.

Non-nuclear wastes are composed of organic and inorganic components. The former decompose into innocuous forms such as carbon dioxide, methane, and water over a period of decades to over a century. In contrast, inorganic materials such as heavy metals do not decay at all. In effect, they have infinite half-lives. Thus, both non-radioactive and radioactive wastes have a component that decays relatively quickly, namely organic wastes and fission products, respectively.

And both wastes have a long-term component, namely inorganic compounds and actinide elements, respectively. The inorganic component of non-nuclear wastes remains toxic forever; the actinide portion of nuclear wastes, such as plutonium-239, decay with time, albeit slowly. Contrary to what one might think , the long lifetimes of nuclear wastes are not unique.

Disposal Methods
So how are non-nuclear wastes of infinitely long hazardous life-time disposed of? Municipal, industrial, and hazardous wastes are all disposed of in the same manner: they are placed into landfills that

consist of large mounds of wastes at ground surface. Modern landfills have plastic and/or clay liners underneath to prevent the escape of leachate, covers on top to prevent the infiltration of rain, and pipes within to collect leachate and gas emissions.

Landfills require large land areas usually near urban centres (Keele Valley landfill, for example, is 380 hectares). Exposed to erosion and precipitation, they must be maintained for long periods of time during which they will emit large quantities of greenhouse and toxic gases. Eventually, no matter what liner technology is used, they will leak toxic materials into the groundwater.

In summary, near-surface municipal and hazardous waste landfills are the accepted method of disposing of non-nuclear wastes, although these wastes are of similar toxicity and lifetimes as nuclear wastes. Now let us see how nuclear wastes are managed.

LOW-LEVEL
RADIOACTIVE WASTE

LLRW is generated by nuclear utilities, medical institutes, and industry, as well as by research organizations and universities. Nuclear utilities generate the following low-level radioactive waste:

• process waste consists of spent ion-exchange resins, filter sludges, evaporator bottoms etc. that arise from cleaning a reactor's moderator, coolant, and storage-pool water. Radionuclides enter into the coolant and moderator by fuel failures and by nuclear activation of corrosion products. Typical of the latter are cobalt-60, iron-55, iron-59, and manganese-54.

• trash including mops, contaminated equipment, used protective clothing, temporary floor coverings, etc. which largely arise from housekeeping operations in those parts of a reactor where radioactivity is present. This category has the largest volume and the lowest activity.

• irradiated components such as control rods, which have reached the end of their useful life or have been damaged.

Typically, a single CANDU reactor in one year produces 12.5 cubic metres of process waste, 250 cubic metres of trash and 25 cubic metres of irradiated components.

Low-level wastes at universities and hospitals arise from the practice of nuclear medicine and from biomedical and other research, and include spent technetium-99 generators, expired vials of radiopharmaceuticals, and contaminated syringes, glassware, gloves, and absorbent pads. Only a few of the nuclides commonly used in nuclear medicine have half lives longer than 28 days; these include iodine-125, cobalt-57 and Yb-169, which are used in very small quantities.

All of the radionuclides employed in nuclear medicine are prepared and processed by radiopharmaceutical manufacturers. The production of some of them, in particular molybdenum-99 and iodine-131, results in relatively large amounts of radioactive waste.

Industry generates a variety of low-level waste including cobalt-60, cesium-137, americium-241, and other sources used in sterilization, radiography, geophysics and a host of other applications.

There are a number of practices used to manage low-level waste, including holding them in storage until they have decayed to innocuous levels (used for radionuclides with short half-lives), incineration, compaction, various liquid treatments such as ultrafiltration and reverse osmosis, and disposal. Extensive research has been conducted to develop and improve these methods, so they can deal with the large spectrum of chemical and physical forms of nuclear waste.

In Canada, no disposal of low-level waste has yet been carried out, where disposal means that wastes are permanently entombed and will require no future maintenance. Instead, long-term, managed storage is used. Internationally, disposal methods include burial in shallow trenches (USA), mined-cavern disposal (Sweden), and near-surface concrete bunkers (France).

In Canada, low-level wastes are treated and stored at two main sites: the Bruce Nuclear Power Development Site of Bruce Power and the Chalk River Laboratories of AECL.

Bruce Nuclear Power Development Site
The central site for treating and storing low-level radioactive waste from Ontario's nuclear stations is located at the Bruce Nuclear Power Development Site. The 8-hectare site includes an incinerator with a nominal capacity of 200-250 cubic metres/month, a mechanical compactor with waste placed into 205-litre steel drums, and a baler

that is similar to the compactor, except that a rectangular container is used to achieve greater stacking efficiency.

All waste at the Bruce site is stored in a solid, retrievable manner in engineered facilities having a 50-year design life. There are four types of storage facilities:

• Reinforced concrete trenches are used for the trash component of LLRW.

• Cylindrical holes are used for process wastes. They are typically 0.7 metres in diameter and 3.5 metres deep and are constructed of concrete with a removable steel liner.

• Three storage buildings are used to store wastes with low radiation fields (less than 10 mSv/hr). Wastes originally in trenches are transferred to the buildings once their radioactivity has decayed sufficiently.

• Quadricells. These are double-walled, above-ground, reinforced-concrete structures used for storing the most radioactive LLRW.

In 1996, a total of 6,304 cubic metre of waste were received (before treatment) with an activity of approximately 8.3×10^{14} becquerels. The entire storage area is underlain by an engineered drainage system which is monitored.

As the Bruce site is for storage only, those wastes that have not decayed to innocuous levels will need to be transferred to a disposal site at some point in the future. A decision has not yet made on how this issue will be dealt with.

Chalk River
AECL has been managing low-level waste from its own research operations and from other waste generators for over 40 years and has also conducted research into classifying, treating and disposing of wastes. Its activities at Chalk River are focussed around the Waste Treatment Centre where liquid wastes are treated by passing them through an evaporator and then through a reverse-osmosis unit; an ultrafiltration unit is also available. The final product is immobilized in bitumen. Compaction and baling are also carried out. The Chalk River site has extensive storage sites which have evolved over a long period. Chalk River receives and manages the low-level waste from nuclear medi-

cine, industry, and research for all of Canada. This function is essential to the operation of nuclear medicine in Canada.

Three major types of storage and disposal facilities are currently on the drawing board. Shallow improved trenches will be used for low-level waste with lifetimes less than 150 years.

A modular above-ground storage facility is currently being designed and constructed that consists of a waste handling building including a super-compactor, a building for storing waste in steel containers, and a building for storing bulk wastes such as contaminated concrete and soil. The objective is to reduce the storage of LLRW in sand trenches.

A shallow mined cavern is proposed for wastes with a hazardous life greater than 500 years. No design work or site selection has yet begun on this concept.

In summary, although more heterogeneous than high-level wastes, low-level wastes are small in quantity in comparison to other non-nuclear wastes produced by society and are contained, rather than emitted into the environment. The relatively small volumes allow innovative disposal methods to be used.

HISTORICAL WASTES

Historical wastes are found primarily in the Port Hope area where Eldorado Nuclear (now Cameco Corporation) initially refined ore to obtain radium for medical purposes and luminous paint. In later years, radium refining was replaced by uranium refining. Wastes from these operations contain the same radioactive elements as in the original ore, although the concentrations and chemical forms may be different.

Port Hope, Ontario
Between 1948 and 1955 wastes were deposited in the Welcome waste management facility located about one kilometre west of Port Hope. The facility contains about 12,000 cubic metres of process waste and 255,000 cubic metres of contaminated soils. A treatment facility collects and treats contaminated ground and surface water.

The Port Granby waste facility, located on the shore of Lake Ontario 16 kilometres west of Port Hope, received wastes between 1955 and 1988. It contains about 204,200 cubic metres of process and other wastes and about 147,000 cubic metres of slightly contaminated soil.

Contamination was spread to locations within Port Hope in a number of ways, including deposition in undesignated sites, use of contaminated materials as fill, spillage from haul vehicles, and wind and water transport from storage sites. The problem of radioactive contamination in the Town was recognized in the mid-1970s and since that time a number of programs have been undertaken to identify contamination, clean up contaminated areas, and consolidate waste into a number of controlled areas. Some contamination also lies at the bottom of the harbour. The amount of waste in Port Hope is about 173,000 cubic metres of waste and 92,000 cubic metres of slightly contaminated soils.

Other Historical Wastes in Canada

About 8,000 cubic metre of marginally contaminated soils are stored in the Toronto suburb of Scarborough as a result of a cleanup conducted in 1996. These wastes arose from a private radium dial-painting facility that operated at the time of the Second World War.

An industrial site with a small amount of historic radioactive wastes (4,000 cubic metre) is located at Surrey, British Columbia. This consists of thorium-contaminated slag from the smelting of imported niobium ore during the 1970s.

In 1992, uranium-contaminated soil and building material were discovered in an unused warehouse in Fort McMurray, Alberta. This was the southern terminus of a long water transportation route from the Port Radium uranium mine in the Northwest Territories that was used in the 1930 to 1950s.

The Low-Level Radioactive Waste Management Office

This office was formed in 1982 and is managed by AECL on behalf of Natural Resources Canada. Its mandate is to deal with the national historic radioactive wastes described above and to provide leadership for managing all other LLRW in Canada. It is currently leading the work to remediate the Port Hope, Scarborough, Surrey, and Fort McMurray sites.

In the early 1980s, efforts were made to find a permanent solution for the historic wastes at Port Hope. Negotiations to find a disposal site broke down, however, and the federal government formed a

Siting Process Task Force in 1986. Its mandate was to develop a process for siting a disposal facility in Ontario for the Port Hope wastes and possibly other wastes. In 1988, the Task Force recommended that siting be based on communities volunteering sites rather than the government dictating locations. In addition, volunteer communities would have considerable involvement in the development of the facility and would also receive compensation.

In early 1989, a Siting Task Force was formed to implement the recommendations. In 1996, following extensive discussions with numerous communities in Ontario, only one, Deep River (near the Chalk River Laboratories), agreed to host the disposal site. The disposal facility was to be at the Chalk River Laboratories site and would consist of a mined cavern. However, discussions between Deep River and the federal government broke down.

With no outside community in Ontario willing to host a disposal facility, an agreement was finally reached with the communities where the waste currently resides. In 2001, the federal government announced that it will allocate $260 million to construct a new disposal facility in Port Hope and to re-engineer and upgrade the existing disposal facilities at Welcome and Port Granby. The process will include an environmental assessment in which possible alternatives will also be assessed.

The Swedish Approach to LLRW Disposal

As Canada has not yet implemented final disposal of LLRW, it is instructive to look at what is being done in other countries. Sweden has developed the most advanced disposal facility in the world, for either nuclear or non-nuclear waste. Because the geologies of Canada and Sweden are similar, this approach could be adapted for use in Canada.

The Swedish Final Repository for LLRW is located on the east coast at the Forsmark Nuclear Power Station about 160 km north of Stockholm. What makes the disposal method unique is, first, that it is located underground and, second, that it is situated under the Baltic Sea. Access to the repository is via two tunnels with entrances on land and which slope gently downward for a distance of one kilometre underneath the Baltic Sea. One of the tunnels is dedicated to waste transportation and is equipped with remotely-controlled vehicles. The repository consists of a large vertical silo (60 metres high and 30 metres in diameter) and four parallel, horizontal caverns. The top of

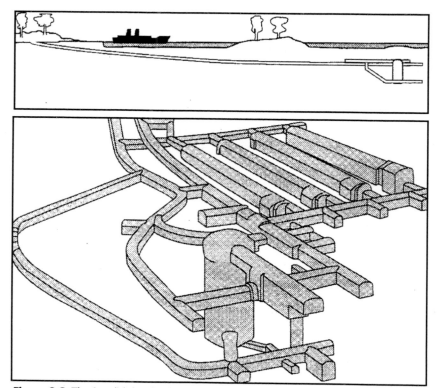

Figure 9-2: The Swedish low-level radioactive waste facility at Forsmark Sweden. As shown in the top panel, entry to the depository is by tunnels that extend out under the Baltic Sea. The waste is deposited in the large silo and in the four horizontal caverns shown in the lower panel.

the repository is about 50 metres below the sea bed and the depth of the Baltic Sea in this area is about 5 metres (Figure 9-2).

The repository started to accept wastes in April 1988. The facility will take all of the low-level radioactive waste generated in Sweden to approximately the year 2010, approximately 90,000 cubic metres. This is less than five per cent of the capacity of a single medium-sized municipal landfill (of which there are thousands in Canada), and illustrates the "small" size of nuclear wastes.

The Swedish Final Repository is like a box within a box within a box within a box. The first barrier is the waste form itself. Before transportation to the repository, the waste is mixed with cement or asphalt and placed in concrete or metal containers.

The silo, which has 0.8-metre-thick reinforced-concrete walls, contains the most highly radioactive materials. The space between the concrete silo and the rock wall is filled with bentonite clay, which expands when it becomes wet. As each layer of waste is emplaced, it is grouted permanently into place with concrete. Thus the engineered barriers consist of the immobilized waste form, grout, the silo wall, and the bentonite layer. Similar barrier systems are used in the four horizontal caverns. The surrounding rock mass provides an additional barrier.

The wastes inside the Swedish Final Repository will decay to about the same level of radioactivity as the surrounding rocks in about 500 years. Less than 10% of the radioactivity will remain after one hundred years. This is about the same length of time as it takes for organic wastes to decompose in a non-nuclear municipal or industrial landfill.

The Swedish repository utilizes only minimal land surface and places no burden on future generations.

In summary, although more heterogeneous than high-level wastes, LLRW are small in quantity in comparison to non-nuclear wastes and are all contained, rather than emitted into the environment. The relatively small volumes allow innovative treatment and disposal methods to be used.

Chapter Ten

High Level Nuclear Waste

Looking at modern nuclear reactors with their valves, refuelling machines, coolant pumps, computer controls, temperature and neutron-flux monitors, and thousands of other precision-made components, it is hard to imagine that nuclear fission could just occur spontaneously in nature. But this is exactly what happened long, long before humans attained that capability.

In 1972, French scientists discovered that at least 17 natural nuclear reactors had operated in the rich Oklo uranium deposit in Gabon in west Africa. This was possible because, due to the natural decay of radioactive elements, their concentrations were much higher in the past. In particular, 1.8 billion years ago, uranium-235 (half-life of 710 million years and present concentration of 0.7%) had a concentration of about 3.7%. This is similar to the concentration used, after enrichment, in today's light-water reactors. Groundwater flowing through the deposit acted as a neutron moderator so that uranium fission reactions started spontaneously and continued for hundreds of thousands of years. Scientists have determined that each of the reactors operated at about 20 kilowatts thermal power.

Extensive scientific studies have been conducted at the Oklo site because the natural reactors form a valuable laboratory for studying how nature has "disposed" of radioactive waste from nuclear reactors. The radioactive products, i.e., wastes, that were created by the reactors (a total of about 5.4 tonnes of fission products and 1.5 tonnes of plutonium) have long ago decayed to stable elements. Close study has shown that the stable daughter products have not migrated from the site, and have remained remarkably immobile — a surprising result given the abundant groundwater and the shallow depth of the ore body, not to mention that the "wastes" were not immobilized into a solid, leach-resistant form, as is proposed for today's nuclear waste. It appears that the clays and bitumen present at the site played an important role in containing the waste. Nature has given us a lesson on how nuclear wastes can be safely contained without harming the environment.

Similarly encouraging results have been obtained from research on uranium ore bodies and from extensive research programs concerning high-level waste disposal throughout the world.

NUCLEAR FUEL WASTES

In Canada, high-level radioactive waste consists of spent fuel which has been irradiated in a reactor until about 67% of the uranium-235 is used up and power can no longer be produced effectively. This takes about one and a half years in CANDU reactors. The spent fuel is sometimes referred to as used fuel or nuclear fuel waste.

Spent fuel contains two types of radioactive nuclides that have been created during the time in the reactor: fission products and **actinides**. Fission products result from uranium-235 atoms splitting when hit by neutrons. Since the uranium atom does not always split in exactly the same way, several dozen different isotopes of approximately half the atomic weight of uranium are formed. These have relatively short half-lives, ranging from seconds to several tens of years. Some of the more prominent fission products are strontium-90, cesium-137, and krypton-85. Fission products generate large amounts of radiation and heat so fuel bundles must be handled remotely and must be shielded and cooled when removed from the reactor.

The second type consists of nuclides that absorb one or more neutrons but do not fission. Instead, they transform into nuclides which have slightly greater atomic weight than uranium, called

Table 10-1
CANDU FUEL BUNDLE COMPOSITION

Isotope	Fresh	Removed from Reactor
uranium-238	99.30%	98.70%
uranium-235	0.70%	0.23%
fission products	-	0.80%
plutonium-239	-	0.27%

actinides (or **transuranics**) such as plutonium-239, americium-241, and neptunium-237. In general the actinides have very low activity, but very long half-lives. Plutonium-239, for example, has a half-life of about 24,100 years.

The composition of a typical CANDU fuel bundle before and after being in the reactor is shown in Table 10-1. The major change is the transformation of about two-thirds of the uranium-235 to fission products. In addition, there is an intermediate reaction in which a small amount (less than one percent) of the uranium-238 absorbs a neutron and transforms to plutonium-239, of which just over half fissions to produce additional fission products. About 30% of the energy derived from the fuel bundle is derived from these plutonium fissions.

Given the enormous energy created by the fuel bundle, it is surprising how little of the material inside the fuel actually changes. The spent bundle looks identical to the original bundle, and only about 1.1% of the material inside has been modified. This is dramatically different from coal, oil and other fuels that undergo a complete physical transformation during their combustion process.

It should be noted that the fissile nuclides, plutonium-239 and uranium-235, together form about 0.5% of the spent fuel, which is only a little less than the original fissile content (0.7% of uranium-235). For this reason, the spent fuel is not necessarily a waste, as these fissile materials could be extracted by reprocessing and used to generate more power. The various fuel cycles in which this fissionable material could be used are discussed in Chapter 14. Liquid wastes generated by such reprocessing are high-level wastes. They could be solidified into glass or ceramic blocks prior to long-term storage or disposal. Reprocessing is not being performed in Canada, nor is it planned.

Ontario Power Generation

Figure 10-1: Pool storage bay for spent fuel. Spent-fuel bundles are placed into baskets and racks with about four metres of water above the top of the fuel racks. The water, which emits a distinctive blue glow from Cerenkov radiation, shields station personnel against radiation.

Spent fuel has two important characteristics. First, it is very small in volume compared to wastes created by many other industries or by burning coal for energy. Second, the waste is contained; it is not emitted into the environment.

INTERIM STORAGE

A 600 MW CANDU reactor produces about 100 tonnes (about 4,200 bundles) of spent fuel per year. The fuel bundles are removed from the reactor core by remote control and placed into deep water-filled pools located at the reactor sites (Figure 10-1), which provide cooling as well as shielding. All the spent fuel produced in Canada up to 1990 would fill one Olympic-size swimming pool.

After the bundles have been in the pool for about six years they can be transferred to dry storage. In Canada, large concrete structures with one-metre-thick walls have been developed for outdoor storage. These structures, called canisters, are being used to store spent fuel from the closed Douglas Point reactor as well as from the Point Lepreau and Gentilly-2 generating stations (see Figure 10-2). In the US, steel as well as concrete canisters are being developed. The water pool and dry storage canisters can store spent fuel for periods exceeding 50 years.

In Canada, one option that is being considered is to move the used fuel from reactor swimming pools to a central location for long-term storage, possibly up to 50 years or longer, prior to it be placed into a final disposal repository.

Hydro Quebec

Figure 10-2: The dry used-fuel storage facility at Hydro Quebec's Gentilly nuclear station. The concrete structures can safely store fuel for decades. The overhead crane system allows remote handling of the radioactive bundles.

In the USA, such a facility, called the monitored retrievable storage facility, is a key part of their fuel cycle strategy because many of their reactors will run out of on-site pool storage capacity before a permanent disposal facility is ready. Although the monitored retrievable storage facility was scheduled to begin accepting spent fuel in 1998, the siting of the facility has run into public opposition and is several years behind schedule.

In Sweden, an underground central spent-fuel storage facility came into operation in 1985. The facility, known as CLAB, will ensure all spent fuel from the Swedish nuclear power program can be safely stored for about 40 years until a permanent disposal facility is ready. The facility uses storage pools much like at a reactor site, except that they are located underground in the granitic rocks of the Baltic Shield.

PERMANENT DISPOSAL

Those countries operating nuclear power reactors are conducting extensive studies on how high-level radioactive wastes should be disposed. They agree that these wastes should be placed into a solid leach-resistant form and then buried deep in a stable geologic formation. At present, no country has yet constructed a repository, although considerable research is being done on a variety of different geologies. Belgium, for example, is studying clay formations. The United States is investigating the suitability of the volcanic tuffs of Yucca Mountain in Nevada. Finland is the closest to implementing disposal of high-level nuclear wastes. In 2001, the Finnish parliament approved the plan to build a repository in the crystalline Precambrian rocks of southwest Finland near the nuclear power plant at Okiluoto.

At this time there is no urgency to build a permanent disposal facility for spent nuclear fuel in Canada because dry storage facilities can provide safe storage for many decades. Nevertheless, it was deemed necessary to demonstrate the feasibility of permanent disposal.

Therefore, an extensive research program was launched in 1975 to develop a concept for disposing of high-level nuclear wastes deep underground. Known officially as the Canadian Nuclear Fuel Waste Disposal Program, its focus was on the rocks of the Precambrian Canadian Shield. Led by AECL, the program included the Geological Survey of Canada, Environment Canada, other federal departments, universities and the private sector. The program conducted research until about 1998, when a formal review of the concept was completed (see below).

A broad range of disposal options have been considered by AECL and the international community, including exotic solutions such as shooting nuclear wastes into space and burying them in polar ice caps or below the seabed. Although some of these solutions are theoretically possible, there are many practical challenges. Shooting wastes into space, for example, is prohibitively expensive and also poses risks, as witnessed by the tragic explosion of the space shuttle Challenger in January 1986. International treaties exclude Canada from disposing of wastes in international territories including the oceans and Antarctica.

Given Canada's geography and geology, the stable ancient rocks of the Precambrian Shield were selected as the most suitable host. This decision has been supported by three independent review groups. The key to safety is placing the wastes deep underground. This simple, yet crucial, step removes the wastes from the forces of erosion and caprices of human intrusion; it also preserves surface land space for other beneficial uses for the growing population.

The only possible method by which the nuclear wastes could become harmful to humans and the environment is if they were to leach out of the canisters in which they are entombed, and then be carried by groundwater to the earth's surface in concentrations that are sufficient to cause damage.

To understand why this is virtually impossible, we need to understand the special character of the Canadian Shield. Rocks of the Precambrian Canadian Shield rocks are amongst the oldest and most stable in the world. Unlike geologic areas with limestone rocks that are soluble, cavities do not exist in this type of rock. In hundreds of kilometres of drilling in granitic rocks of the Shield, only cracks (no cavities) have been observed, of which the largest are only a few centimetres in width. No evidence has been presented that cavities exist in the plutonic rocks of the Shield of dimensions similar to the proposed waste canisters and repository rooms. This is in spite of the very long time that ground water has been present in these rocks (a large portion of the Canadian Shield is about 2,500 million years in age).

For nuclear waste disposal, the period of concern is 10,000 years, a small length of time compared to the age of the Shield. Even if the natural leaching process were increased a thousand fold by the heat generated by the waste, simple calculations show that only trivially small openings would be leached over 10,000 years. The reasons are that, first, granitic rock is composed of essentially insoluble silicate minerals, similar to glass, and, second, although fractures exist, water resides mostly at the surface because the permeability and porosity of these cracks decrease with depth. Little groundwater exists at the depth of the disposal vaults and the water which is present is very old and very slow moving.

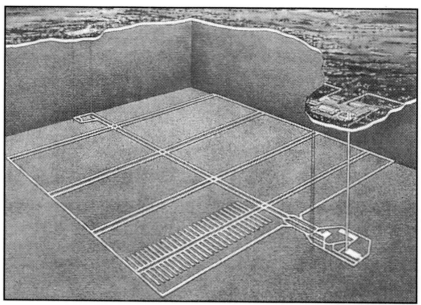

AECL

Figure 10-3: Conceptual design of a spent-fuel disposal facility. Fuel bundles sealed in appropriate containers would be deposited in disposal rooms, which would then be back-filled and sealed.

Thus, as long as the job is done carefully and thoroughly, granitic formations should safely entomb nuclear wastes for millions of years. A suitable geologic formation must be selected, wastes must be immobilized in solid leach-resistant form, and the tunnels and shafts of the repository must be properly sealed once it is full.

The repository in AECL's disposal concept consists of a grid of tunnels at a depth of about 500 to 1000 metres. Waste containers would be placed in vertical boreholes drilled into the floors of the tunnels. The boreholes would be plugged with materials such as bentonite clay. Once the waste is emplaced, the tunnels, shafts and exploratory boreholes would be sealed and grouted. The conceptual repository is shown in Figure 10-3. An alternate design was also developed where waste containers are placed horizontally in the tunnels and surrounded by bentonite clay.

A key aspect is that the waste would be placed into a very inert, stable, solid form before emplacement underground. Research indicated that titanium or copper containers would have the necessary corrosion-resistant properties. Glass beads would be compacted into the spaces between the fuel and the container shell to provide internal support against underground pressures. These containers, which are cylindrical in shape and contain 72 fuel bundles, have been designed so no waste would be released from them for at least 500 years.

Research was also conducted into the materials and methods for sealing the repository and associated shafts and boreholes. A mixture of clay and sand packed around the waste containers would limit groundwater access and would trap radioactive material that might escape. On contact with water, the clay would swell and provide a seal around the containers. The geologic strata surrounding the repository would provide the main barrier. Extensive research was conducted into methods of characterizing a rock mass and understanding how ground water moves through it.

The main barriers to release of radioactivity would consist of:

- the waste form, i.e. ceramic uranium dioxide fuel pellets
- the waste container
- buffer material surrounding the container
- the geologic formation.

Considerable research was conducted to study the combined effects of heat, stress and chemistry on the groundwater system and its capability to breach containers and leach and transport radionuclides, the only possible method by which the radionuclides might be able to escape.

The main way of demonstrating the long-term safety of this disposal concept was through a sophisticated computer program called SYVAC (Systems Variability Code), which models the movement of radionuclides as they are leached from the waste containers and carried by groundwater from the repository through the rock mass to food chains and thence to human beings. These analyses, known as a risk assessment, are complex, as all the pathways by which radionuclides may reach humans must be identified and modelled in mathematical form. It is important that the physical, chemical and biological processes involved be well understood and that data describing them be obtained from the site under question.

Much of the data for the computer analyses was obtained from the Underground Research Laboratory located in the Lac du Bonnet batholith, a large granite rock mass near AECL's Whiteshell Laboratories in Manitoba. A shaft was sunk to a depth of 445 metres in

Figure 10-4: The Underground Research Laboratory near Pinawa, Manitoba. Engineers are drilling test holes to measure the properties of the rock.

previously unmined rock, and a number of galleries and rooms were excavated in which various experiments have been and continue to be conducted. The experiments studied the effects of heat on rock, the transport of radionuclides through rock fractures, methods for sealing rooms and boreholes, methods of mining and blasting that minimize damage to the surrounding rock, and much more. Figure 10-4 shows tests being performed in the Underground Research Laboratory.

The risk assessment provided quantitative predictions of potential doses to humans, information which was used to assess the safety of the system and also to provide a reference against which future performance could be measured. The results presented by AECL showed that the concept is safe and would meet all regulatory requirements.

Considerable international review and input have gone into the Canadian program. For example, Japan, the USA, Sweden and France have joined in cooperative research programs at the Underground Research Laboratory. The program also received extensive review in Canada. The Technical Advisory Committee was formed by AECL in 1980 to provide independent, high-quality review of the program. The Technical Advisory Committee was composed of distinguished scientists and engineers who have been nominated by various scientific organizations and who have no direct links with AECL or the nuclear utilities. The Committee produced an annual report providing a technical review of the disposal program.

REGULATION

The Canadian Nuclear Safety Commission is the regulatory agency with responsibility for licensing all nuclear matters in Canada including the disposal of spent fuel. It has issued a series of regulatory guides and is monitoring the national high-level waste disposal research program. The CNSC requires that the burden on future generations be minimized by:

• selecting disposal options that do not rely on long-term institutional controls;

• ensuring that there are no predicted future risks to human health and the environment that would not be currently accepted;

• ensuring that the radiological risk does not exceed one in a million serious health effects per year.

The regulations imposed on nuclear wastes are far more stringent than those for non-nuclear wastes. A main difference is that computer-based risk assessment is mandatory for licensing a nuclear waste disposal facility but is not required for licensing hazardous-waste disposal facilities. Another difference lies in the consideration of future generations. Nuclear waste regulations require that disposal methods must not rely on long-term institutional controls and mathematical risk assessment is to be performed up to 10,000 years into the future.

In contrast, municipal landfills in Ontario, for example, do not have to predict any future impacts. There are no specific post-closure criteria although guidance manuals describe requirements for a surface cap and long-term monitoring. The Ontario Ministry of Environment's approval is required if a landfill site is to be used for other purposes within 25 years of its closure. There is also a requirement to maintain engineered works such as leachate collection systems "as long as needed". No discussion is provided as what might constitute "as long as needed".

The situation is the same for hazardous wastes. The Canadian Council of Ministers of the Environment guidance document makes little mention of the post-closure period other than to state that "hazardous wastes can retain their harmful properties over a long period of time, perhaps for centuries." This body recommends that a post-closure program be implemented that includes insurance, main tenance, contingency plans, and monitoring. Depending on site conditions, it states that this post-closure period "could well exceed 100 years."

Unlike nuclear waste disposal, no systematic thinking has gone into the long-term implications of municipal and hazardous landfills. There are no definitions of the post-closure period, nor has there been consideration of long-term issues such as ice-ages, seismic activity, or future human intrusion, which are routinely assessed for nuclear wastes.

Furthermore, because the facilities are located at the surface, perpetual on-going maintenance will be required. Put bluntly, municipal, industrial, and hazardous landfills only address the short term and are designed and regulated to give the problem to future generations.

Canadian Environmental Assessment Agency Panel Review

AECL's disposal concept for nuclear fuel waste was referred for review under the federal environmental assessment review process in 1988. A year later, a panel was appointed, composed of technical experts in earth sciences, biology, mining, and nuclear medicine. Non-technical representatives included a former president of the World Council of Churches and a member of the First Nations. The Panel was supported by a Scientific Review Group of scientists and engineers.

After ten years and following extensive public hearings that included 561 written submissions, the Panel presented its results in 1998. The panel stated that the safety of the concept had been adequately demonstrated from a technical point of view, but not from a social perspective. As the concept did not have broad public support, it therefore was not acceptable as Canada's approach for disposing of nuclear waste.

The panel made a number of recommendations for a program that it felt would ensure broad public support, including:
• an Aboriginal participation process,

• a comprehensive public participation process,

• an ethical and social assessment framework, and

• an arms-length agency to manage nuclear waste disposal, separate from AECL and the nuclear utilities.

The panel recommended that the search for a specific disposal site should not be pursued until these recommendations are implemented and broad public acceptance of the nuclear waste disposal concept is achieved.

In December 1998, the government of Canada issued its response to the Panel's recommendations. After receiving input from various stakeholders and the public, the government introduced nuclear fuel waste legislation in April 2001, which requires the nuclear utilities to establish a separate waste management organization to manage the disposal of nuclear fuel waste. A dedicated fund is to be established by the nuclear utilities to cover the cost of the long-term management and disposal of nuclear fuel wastes. A review and approval mechanism will provide access to the fund. A reporting relationship is to be established between the federal government and the new waste management organization. The new waste management agency is to assess disposal alternatives including the option of long-term centralized storage, and is to establish a comprehensive public participation program.

SUMMARY

There is a perception that nuclear wastes are uniquely toxic, extremely long-lived, and that scientists cannot find methods for their disposal. A comparison shows that many non-nuclear wastes have similar toxicity, are even more long lived (infinite), and are much greater in volume.

In Canada, non-nuclear hazardous wastes are disposed of in surface landfills. In comparison, it is proposed to dispose of encapsulated spent fuel into a single facility deep inside a carefully-selected, ancient geologic formation.

It appears that the method proposed for nuclear waste disposal will provide containment and safety for a very long time, and is consistent with the principle of sustainable development.

Chapter Eleven

Nuclear Medicine: The Gift of Life

The practice of medicine has been profoundly transformed by nuclear technology. Where exploratory surgery was previously necessary for doctors to peer inside the body and diagnose problems, now a host of radioactive tracers are available to perform this investigation. Where surgery was previously necessary to remove cancer tumours, now they are destroyed painlessly by gamma rays. Today, nuclear techniques are so prevalent that one out of three Canadians entering a hospital undergoes a nuclear procedure of some kind. A new medical branch, nuclear medicine, has evolved over the past four decades, and Canada has been at the forefront in developing and using the high-technology nuclear tools that made it possible. In 1969, for example, Quebec doctors were the first in the world to recognize nuclear medicine as a separate specialty.

DIAGNOSTIC NUCLEAR MEDICINE

Nuclear medicine is composed of two distinct parts: diagnosis (discovering what the disease is) and therapy (treating and curing it). The large majority of nuclear medicine procedures are diagnostic rather than treatment.

X-rays

X-rays are commonly used in both medical and dental practice, and everyone has had X-rays taken of their teeth or perhaps of a broken arm or leg. The shadowy images that appear on the photographic film reveal dental cavities as well as bone fractures highlighted against the denser bone. Subtle differences in other organs can also be detected and provide clues about the presence of disease.

X-rays are photons resulting from the collisions of electrons with atoms. Strictly speaking X-rays are not a nuclear but an atomic phenomenon, although they are physically the same as gamma rays. The difference is that X-rays originate from changes in the energy levels of atomic electrons and gamma rays from the energy levels of nuclei. Figure 11-1 shows the first X-ray picture ever taken.

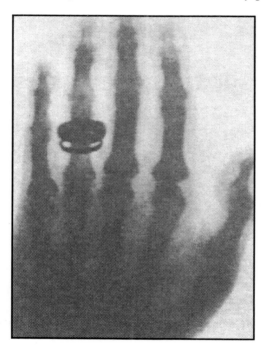

Figure 11-1: The first X-ray picture ever published. It was taken by Wilhelm Roentgen of Germany in 1896 and shows his wife's hand. For this discovery Roentgen received the first Nobel Prize in Physics in 1901.

About 31 million X-rays are taken each year in Canada, just over one per person per year. Chest X-rays account for 26%, and shoulder, pelvis and limbs account for 22%. Dental X-rays are also common, accounting for about 14%. Radiation doses range from 0.02 mSv for a dental X-ray to about 0.07 mSv for a chest X-ray. X-rays are also emitted by television screens and computer monitors although the doses to individuals are small.

Radioactive Tracers

Diagnostic nuclear medicine is possible because the nuclear decay process emits particles that can be detected from a distance. An appropriately radioactive-tagged drug, called a **radiopharmaceutical**, is sent into the body either by injection or orally. The drug is selected so it will be absorbed preferentially by the specific organ which is to be examined. Today some 20 to 50 radiopharmaceuticals are in routine use for diagnosis, most of which are organic in nature. The path of the radiopharmaceutical is followed by scanning the relevant organ with a special instrument called a gamma camera, also called an Anger camera, which detects even minute amounts of gamma radiation. This procedure avoids the previous practice of exploratory surgery, which can be very traumatic. Nuclear medicine is making the scalpel obsolete for diagnosis.

It should be noted that radioactive tracers provide information on how organs function. In this way they complement X-rays, which show the shape of organs.

Although there are more than 1,500 known radionuclides, only a few are suitable for medical diagnosis. The primary requirement is to minimize the patient's exposure to radiation dose by using gamma rays with just enough energy to be detectable outside the patient's body and with radiation that lasts just long enough, that is with a short half-life, to allow completion of the diagnostic tests. These radionuclides are mixed with chemical compounds that are attracted to the organs of interest. Radioisotopes such as technetium-99m, iodine-131, and thallium-201 are used to study the heart, brain, bones, lung, thyroid, liver and kidneys (see Table 11-1).

Table 11-1	ISOTOPES USED IN MEDICAL DIAGNOSIS
Isotope	Organ System(s)
Technetium-99	brain, lung, bones, heart, infection, tumour
Thallium-201	heart
Iodine-131	thyroid
Iodine-125	blood, radioimmunoassay, prostate
Iodine-123	thyroid, heart, brain
Xenon-133	lungs
Gallium-67	soft tissues
Indium-111	infection, inflammation
Strontium-82*	heart

*produces rubidium-82

Approximately 15 to 20 million diagnostic tests are performed annually using nuclear tracers.

Tc-99m is the most important and widely used isotope in nuclear medicine. It has a low-energy gamma and short half-life (six hours) so it does not stay in the body very long. Technetium-99m was discovered in 1937 and came into general usage in 1973 when Chalk River began regular production of its parent, molybdenum-99.

THE TRAVELS OF MOLY

Molybdenum is the parent of technetium, the most important and widely used nuclear tracer in medicine. Its delivery to hospitals around the world is a formidable exercise in logistics, since molybdenum has a half-life of only 66 hours. Let us look at how it works.

Molybdenum is produced by irradiating an uranium-aluminum alloy for about two weeks in the NRU reactor at Chalk River (two new MAPLE reactors will take over this job beginning in 2002). Molybdenum-99 is created as one of the fission products. The irradiated uranium-alloy coupons are removed from the reactor in the afternoon and are processed by remote control in heavily shielded hot cells. The coupons are dissolved and the resulting solution is passed through an ion-exchange column to extract the molybdenum-99.

The molybdenum-99 is rushed to Ottawa, about a two-hour drive, where MDS Nordion technicians work overnight to do the final processing and packaging. Early next morning, the molybdenum processing and testing are complete and the product is packaged for shipment. The molybdenum is then delivered to the Ottawa airport and flown by chartered air plane to the US radiopharmaceutical producer that uses the molybdenum-99 to make technetium-99m generators. Shipments to Europe, Japan, and South America are made by scheduled cargo flights. Hospitals are supplied with portable lead-lined "generators" from which technetium-99 is drawn off, or 'milked', as needed.

Tc-99m is flexible chemically and can be formulated into numerous specific compounds for a variety of functions. For example, when combined with albumin particles and injected intravenously, it is trapped in the blood vessels of the lungs, helping to identify areas with decreased or absent blood flow. Linked to a phosphate molecule, it will concentrate in bone tissue at a rate proportional to blood flow and bone production level. More than twenty different technetium compounds are used, and it is estimated that world-wide at least five million people undergo Tc-99m diagnostic procedures annually. Canada is the world's leading supplier of Mo-99.

Because of its very short half-life, technetium itself cannot be shipped to hospitals. Instead, its parent molybdenum-99 is manufactured and shipped across the world. Figure 11-2 shows how molybdenum is packaged for shipping. Most of the world's supply is specially made in the NRU reactor (now being replaced by two MAPLE reactors). The sidebar describes the highly coordinated process of getting this important radioisotope to hospitals around the world in a timely fashion.

Figure 11-2 Container system for shipping radioactive molybdenum. Note that several containers are layered one within the other for maximum safety. About a million shipments of radioisotopes are made each year in Canada using such containers.

The workhorse of most nuclear medicine departments is the gamma, or Anger, camera (see Figure 11-3). The camera is composed of four main components: a collimator, a scintillation crystal, a light detection system, and a computer processor. As shown in Figure 11-4, gamma rays are emitted from an organ that has been 'tagged' with an appropriate radionuclide. The collimator is made of lead and contains many parallel holes that ensure that only those gamma rays travelling parallel to the holes pass through. The crystal is composed of pure sodium iodide that has the special property that when a gamma ray passes through, it scintillates, that is, it flashes a blue light whose intensity is proportional to the energy of the gamma ray. Behind the crystal is a layer of photomultiplier tubes that convert the light flash to a measurable electric voltage that is proportional to the size of the light flash and thus the energy of the gamma ray. Finally the data is processed by computer to produce an image of the activity in the

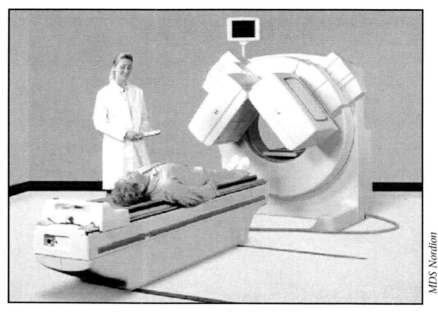

MDS Nordion

Figure 11-3: The gamma camera produces images by detecting the gamma rays emitted from radioisotopes contained in various chemicals put into the body. Since the camera has very high sensitivity, tiny quantities of radioisotopes give excellent images.

organ. By positioning the camera in the right place, the source of the gamma rays can be determined. Computer techniques are used to form an image showing where the radioisotope is located. Such images are then used to deduce how that organ is functioning.

A drawback of the gamma camera is that it only gives a two-dimensional picture. Two other methods, Positron Emission Tomography (PET) and Single Photon Emission Computerized Tomography (SPECT), have been developed that allow three-dimensional views.

Nuclear diagnosis techniques are not particularly useful in demonstrating anatomy, but are very powerful in demonstrating how various organs function.

Nuclear medicine touches almost everyone. We all have either had a nuclear procedure ourselves or have a relative or close friend that has. For example, roughly 3% of the population of Canada, almost a million people, suffer from coronary artery disease. Most of them will be diagnosed using radiopharmaceuticals and nuclear techniques, and this is but one of hundreds of diagnostic nuclear procedures.

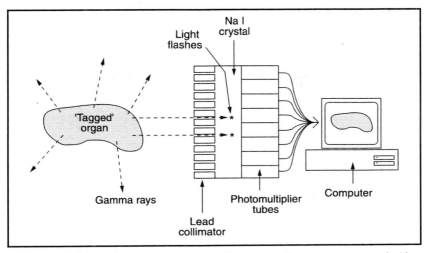

Figure 11-4: Schematic of a gamma camera. Gamma rays from an organ tagged with a radiochemical cause light flashes in a sodium iodide crystal that are detected by photomultipliers that send the elctronic pulses to a computer for analysis and display.

In Vitro Methods

In vitro methods involve taking a sample of blood or tissue from a patient and performing analyses in a laboratory with no radiation dose to the patient. These procedures are called *in vitro* (Latin for in glass, i.e., in a test tube) compared to *in vivo* (in a living person) techniques. A method called radioimmunoassay uses radioisotopes to detect evidence of cancer by revealing the presence of antigens, or tumour markers, which indicate cancer may be present. This method is millions of times more accurate than any previous technique. A radioactively labelled substance (an antibody, for example) is mixed with a sample of blood that has been taken from a patient. By measuring the amount of labelled substance that reacts with the sample, the amount of that substance can be calculated very accurately, because the radioactively tagged substance competes with the non-tagged substance in the sample for a known number of binding sites. Radioimmunoassay can also be used to detect infectious diseases and measure minute quantities of hormones, vitamins, and drugs in bodily fluids.

Positron Emission Tomography (PET)

Positron Emission Tomography is used to study subtle chemical changes and metabolism in the brain and other organs. A radiochemical that emits positrons is taken up by the organ to be studied. A positron is very unstable and immediately is annihilated by interacting with an electron, creating two gamma rays that travel in opposite directions. A ring of gamma-ray detectors is placed around the part of the body containing the organ under study. When a pair of detectors on opposite sides of the ring each identify an annihilation gamma ray at the same instant, then a positron decay must have occurred on the line joining the two detectors. A ring of detectors is used that is capable of detecting many annihilation events. If a sufficient number of lines of interaction are identified, then an image of the distribution of radioactivity can be constructed.

As an example, PET scans can view the distribution of fluorine-18 in a plane taken through the head (see Figure 11-5). By measuring several planes, a three-dimensional picture can be obtained. The extent to which different regions of the brain are functioning and the response of brain tissue to external stimuli can be tested in both normal and diseased states. For example, an epileptic fit suffered by a child shows up as a dramatic increase in activity in a defined region of the brain. For this reason, PET scans are used to study brain disturbances such as epilepsy, brain tumours, strokes, and can even detect how the mind works.

Figure 11-5: Imaging a human brain using PET. A small amount of a dopamine precursor labelled with fluorine-18 is injected into the patient. The pattern of the gamme rays detected indicates the distribution of the dopamine, a key chemical involved in brain function. By studying these patterns physicians can diagnose a variety of brain disorders.

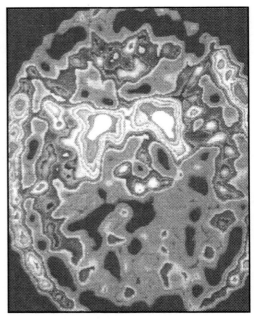

Hamilton Health Sciences

At the McMaster University Hospital in Hamilton, Ontario, which has one of the major nuclear-medicine departments in Canada, radioisotopes that emit a positron (fluorine-18, carbon-11, nitrogen-13, oxygen-15) are made in a small cyclotron located at the hospital dedicated to producing the neutron-deficient radioisotopes used in PET scans.

Other Nuclear Techniques

The Single Photon Emission Computerized Tomography (**SPECT**) camera has gained widespread acceptance over the last few years for use in routine clinical imaging. The SPECT camera provides views of thin slices of an organ rather than the entire organ. A detector head is used that rotates around the patient through 360 degrees collecting images that are stored in a computer. Computer processing techniques are used to reconstruct the slices into three-dimensional images of the organ.

Another instrument, the dual-photon gamma detector, calculates bone densities with great accuracy aiding in the understanding of osteoporosis. The instrument directs gamma rays of two different energies through the bones of interest. By measuring the difference in gamma-ray attenuation, doctors can determine bone density. By repeating these measurements over time, doctors can determine how bone density changes with aging.

Canadian studies have shown that strontium-89 can be used to ease pain in patients that have bone cancer. Not only does the patient feel better, but the treatment costs about $20,000 less than conventional treatment.

Radionuclide Manufacture

MDS Nordion in Ottawa, Ontario, is the world's leading producer of radioisotopes for medical and industrial uses. It supplies about two-thirds of the world demand for reactor-produced isotopes as well as substantial amounts of cyclotron isotopes. Isotopes with an excess of neutrons (carbon-14, chlorine-36, nickel-63, molybdenum-99, iodine-125, iodine-131, xenon-133, and others) are produced in the NRU reactor at Chalk River Laboratories.

With the NRU reactor reaching the end of its life, two new 10 MWt pool-type reactors called **MAPLE** 1 and 2 (Multipurpose Applied Physics Lattice Experiment) are being commissioned at Chalk River. These reactors, owned by MDS Nordion and operated by AECL, will be dedicated to making medical radioisotopes including molybdenum-99, iodine-131, xenon-133, and iodine-125. Each reactor has a compact core of low-enriched uranium, surrounded by a vessel of heavy water, at the bottom of a light-water pool. The MAPLE project also includes a processing facility where the isotopes will be extracted and packaged for shipment to MDS Nordion in Ottawa. The processing facility has been designed to reduce the volume of radioactive waste produced.

Thallium-201, widely used in heart studies, is the most significant isotope produced in a cyclotron. A cyclotron, also called a particle accelerator, uses an alternating electric field to accelerate charged particles such as protons to a very high speed. They are guided around a circular track by magnets and then shot at specially prepared targets. The collision produces radioactive substances that are deficient in neutrons including cobalt-57, gallium-67, indium-111, iodine-123, strontium-82, as well as thallium-201. MDS Nordion operates two cyclotrons that are dedicated to radioisotope production and also has access to the large TRIUMF cyclotron at the University of British Columbia (see Chapter 16).

The impact of these Canadian-produced medical radioisotopes is enormous. About 12 million medical procedures are conducted each year around the world using medical isotopes supplied by MDS Nordion.

Harold Johns (1905-1987): started his career as a professor of physics at the University of Alberta. In 1945, he moved to the University of Saskatchewan, at the same working at the Saskatchewan Cancer Commission. It was there that he designed and built Canada's first cobalt therapy machine, one of the first in the world. From 1956 until his retirement in 1980 he was a professor at the University of Toronto. He was inducted into the Canadian Medical Hall of Fame in 1968 and was awarded the Order of Canada in 1976.

AECL

THERAPEUTIC NUCLEAR MEDICINE

Therapeutic nuclear medicine, which deals with treating and curing diseases, differs dramatically from diagnostic nuclear medicine. In diagnostic procedures, radionuclides are placed inside the body and radiation doses are small. In contrast, the doses administered in radiotherapy are delivered by sealed sources kept outside the body; the doses delivered are large.

Cancer is a deadly disease in modern society accounting for 20 to 25% of natural deaths in North America. Furthermore, the incidences of some forms of cancer such as lung cancer are increasing by about 0.5% per year. Needless to say, a cure for cancer is a much sought-after goal.

As long ago as the 1890s, X-rays were used to treat breast cancer. Today, radiation is used to kill cancer tumours by carefully focussing gamma-ray beams generated by either cobalt-60 or accelerators onto the tumour. Cancerous cells are more sensitive to radiation than healthy cells. The beam is aimed at the tumour from many different directions so that nearby healthy cells receive much less radiation than the tumour. It is important that gamma beams of the appropriate energy be used, and that they are focused very accurately.

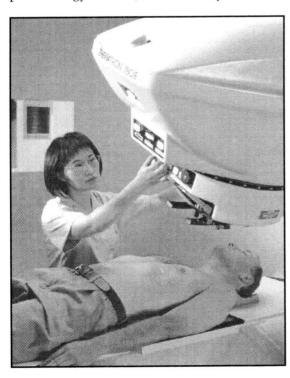

Figure 11-6: A cobalt cancer-therapy machine. The cobalt is contained in the heavily-shielded 'head' of the machine. The cobalt source is exposed by opening a shutter, and the 'head' is rotated to aim radiation at the tumour from several different directions to maximize exposure to the tumour while minimizing exposure to the surrounding tissue.

MDS Nordion

Canada has been a leader in the development of methods for treating cancer. The first cobalt-60 cancer-therapy machine was invented in Canada and was initially used at Victoria Hospital in London, Ontario, on October 27, 1951 (see Figure 3-5 in Chapter 3). For many years, Canada was the only country that could produce cobalt-60, and today it supplies over 80% of the world's demand.

MDS Nordion has supplied more than 2,500 cobalt therapy machines, over half the number in the world, that are used in hospitals in over 80 countries. These machines deliver over 40,000 treatments each day around the world. MDS Nordion markets three different models of radiation therapy units, as well as a sophisticated computer model for planning and tracking radiation treatment. A modern cobalt-60 cancer therapy machine is shown in Figure 11-6.

Radioactive cobalt used in cancer therapy treatment units requires an activity that is about four times greater than that of the cobalt used in industrial irradiation machines (see Chapter 12). Cobalt for irradiation machines is made in Ontario Power Generation CANDU reactors, whereas the more active cobalt for therapy is "custom" made in AECL's NRU reactor.

Radiation treatment can prolong the lives of patients suffering from cancer for some number of years. Doctors who specialize in a field of medicine called epidemiology have tracked large numbers of such cancer patients and have calculated that cobalt-60 therapy machines have created approximately 13 million years of additional life.

Another type of radiation therapy known as brachytherapy puts radioactive sources, or "seeds", directly into or near the patient's tumour. This has the benefit of eliminating the dose to skin and healthy tissue between the tumour and the skin. These seeds, which are generally produced using cyclotrons, are introduced into the body through surgically-placed tubes called catheters.

While the role of radiation in the treatment of cancer is well known, less known is its application to treating heart disease. When coronary arteries become blocked one option is to use catheterization techniques to open the blockage by installing a stent, a small tube of wire mesh that holds the artery open. The problem is that the stent itself can become blocked. A promising solution to this dilemma is to use a stent containing a special radioactive material that discourages clotting. This method is now coming into increasing use throughout the world.

The impact of nuclear technology on the practice of medicine is profound. More than one in three people have undergone some form of nuclear medicine treatment, the ability to make accurate diagnoses has improved, exploratory surgery has decreased, and millions of years of life have been saved. Canada has been a world leader in this field.

Chapter Twelve

Nuclear Technology in Industry & Science: Unlimited Potential

The application of nuclear techniques in the industrial and scientific realm is not as dramatic as in medicine, but is even more far-reaching. Almost everything we do or every product we use is of higher quality, cheaper price, or simply offers more choice and variety because of the ingenious use of nuclear technology. As described in this chapter, the uses of nuclear technology are limited only by human imagination.

The ability to use radioactive isotopes in so many ways is a result of three unique nuclear properties, namely:
- radiation can penetrate matter;

- radiation can be detected easily and accurately;

- there are thousands of radionuclides in a wide range of radiation energies and activities; they can be made into almost any physical or chemical form.

IRRADIATION

With the recognition, over 100 years ago, that germs spread diseases, various methods of sterilization have evolved, many of them depending on the use of heat. The potential for using irradiation with gamma rays was recognized in the late 1950s. It permits the sterilization of products such as plastics that are damaged by heat, and is often the only means of sterilizing some pharmaceutical powders, solutions, and ointments. A particularly important feature is that the penetrating character of gamma rays allows medical supplies to be sterilized after they have been packaged and sealed, thus preventing subsequent contamination.

Figure 12-1: A Gamma Cell 220 irradiator manufactured by MDS Nordion of Ottawa, Ontario. These irradiators have many applications in research and medicine. Over 1,100 of Nordion's Gammacell irradiators have been installed worldwide.

MDS Nordion

In 1958, MDS Nordion's predecessor, AECL's Commercial Products Division, developed the Gammacell 220, the first research **irradiator**, which was sold to North Carolina University. Today the Gammacell 220 is widely used by hospitals, blood banks, and laboratories for irradiating small samples of blood for immuno-incompetent or immuno-compromised patients or for research. One of the best ways to reduce the risk of Graft Versus Host Disease in patients with severely weakened immune systems, such as premature babies and bone-marrow transplant recipients, is to use irradiated blood when performing transfusions. The Gammacell 220 is shown in Figure 12-1.

NUCLEAR GOLF BALLS

A sports company in the USA has released "the world's longest legal golf balls". The balls, which are irradiated with gamma rays from a cobalt-60 source, were used to set a world record drive of 367 yards on the fly. The gamma radiation increases cross-linkages between molecules making the golf balls more resilient. The company's line of irradiated tennis strings is also receiving acclaim.

In 1964, the first full-scale commercial irradiator in the world was developed in Canada, and by 2001 there were a total of about 170 irradiation plants in 65 countries around the world. More than 120 were supplied to over 45 countries by the Canadian company, MDS Nordion.

Today, more than 40% of the world's disposable medical products are sterilized with cobalt-60 gamma radiation. Articles such as syringes, needles, sutures, intravenous tubing, gloves, gowns, sponges, and catheters are packaged and then sterilized while inside the packages. Doctors, scientists, and medical technologists have measured, used, and studied nuclear irradiation for decades on equipment and supplies that have been used to treat millions of patients, and they have reached a simple conclusion: it saves lives by preventing contamination.

Research irradiators are used to study the effects of radiation on a wide variety of materials. For example, electronic components that will be used in satellites are tested in research irradiators to see how they respond to the radiation fields encountered in space.

The radioactive cobalt that forms the heart of irradiation units is made in Ontario Power Generation's reactors and encapsulated by MDS Nordion. Some of the stainless-steel adjuster rods used to control the nuclear reaction are replaced with rods made of cobalt-59. These are installed when the reactor is shut down for servicing and are removed after about one year, during which time the cobalt-59 has absorbed neutrons and become cobalt-60. With a half-life of 5.26 years, its radiation intensity decreases by about 1% per month. Canada provides about 80% of the world's supply of cobalt-60.

Food Irradiation

Illness due to contaminated food is "perhaps the most widespread health problem in the contemporary world and an important cause of reduced economic activity", says the United Nation's Joint Expert Committee on Food Safety. And food poisoning is not restricted to third-world countries. In Canada, it is estimated that more than seven million cases of food-borne illness occur each year with a death toll of between 200 and 500, and a cost to the health care system of more than $1 billion annually. Thus, irradiation of food has the potential to alleviate an enormous amount of human suffering.

Gamma rays are sent through the item being irradiated, disrupting organic processes, breaking down microbial cells, such as bacteria, yeasts, and moulds. Parasites and micro-organisms such as trichinella and salmonellae, insects, and their eggs and larvae are killed or made sterile. A commercial food irradiator is shown in Figure 12-2.

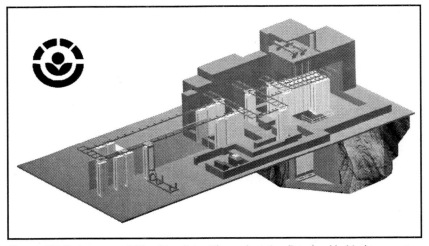

MDS Nordion

Figure 12-2: A commercial food irradiator. These plants irradiate food behind two-metre-thick concrete walls in a bungalow-size building. The cobalt-60 gamma-ray source resides under the concrete building and is raised into position once the food packages are in place. Inset is the radura, an international symbol that identifies food that has been irradiated.

One of the world's highest priorities is to produce enough food to feed the ever-increasing population. Considerable effort is invested in fertilizing farmland, improving crop varieties, providing irrigation, and improving animal husbandry. With some countries reporting up to 40 to 50% post-harvest losses through infestation of staple foods like grains and yams, it is also important to make progress in preserving food once it has been grown. Preservation methods have evolved from sun drying, to salting, smoking, canning, freezing, heating, and the addition of chemicals. Food irradiation is compa-rable to pasteurization or freezing and can slow the ripening of some fruit and inhibit the sprouting of root crops. It preserves food by destroying micro-organisms that cause normal spoilage; for example, poultry that has been irradiated will stay fresh for up to two weeks, compared to three days if not irradiated.

Although food irradiation is relatively new, it has been studied more thoroughly than any other food preservation method. These studies show that irradiation is effective and there are no adverse effects from the consumption of irradiated food. The World Health Organization, the Science Council of Canada, the American Medical Association, and other expert bodies endorse food irradiation. It is important to realize that irradiation is virtually identical to the process used by X-ray machines such as at airport security or the dentist's office. Food irradiation does not make food radioactive, just as dental X-rays do not make your mouth radioactive.

Labelling requirements in Canada and the US are similar, requiring irradiated food packages to bear the international radura symbol as well as one of the two statements: "treated with radiation" or "treated by irradiation". The radura is shown inset in Figure 12-2.

Food irradiation is now used relatively widely, especially in third-world countries. There are about 30 food irradiators worldwide and more than 40 countries have approved the use of gamma irradiation for hundreds of foods, including fresh fruits and vegetables, grains, spices, poultry, and seafood. In 1984, the Food and Agriculture Organization and World Health Organization of the United Nations published a general code regarding food irradiation that states: "The irradiation of foods up to an overall average dose of 10 kilogray introduces no special nutritional or microbiological problems." Based on exhaustive research, this statement provides justification that food can be safely irradiated, and forms the basis for food irradiation legislation in many countries.

US interest in food irradiation has been accelerated by recent media headlines about food poisoning. In 1993 in the western USA, four children died and several hundred became ill after eating hamburgers containing the *E. coli* 0157:H7 micro-organism. In 1997, Hudson Foods recalled 11 million kilograms of red meat including hamburger patties because the meat had been contaminated with harmful *E. coli*. Subsequently, this industry giant filed for bankruptcy. Furthermore, methyl bromide, the most commonly used food pest-control fumigant, has been listed as an ozone depleting substance and is being phased out.

Good progress is being made in implementing food irradiation in the USA. In 1985, the US Food and Drug Administration approved the irradiation of pork to control the parasite that causes trichinosis. In 1990, it approved the irradiation of poultry, and in 1997, the irradiation of fresh and frozen meats including beef, lamb, and pork. In 1991, North America's first commercial irradiator dedicated to food processing was installed near Tampa, Florida, by MDS Nordion. More recently, progress has been made in pasteurizing meat products using high-energy electrons.

Canada is a leader in developing food irradiation technology. A full-scale food irradiation facility was established at the Canadian Irradiation Centre at Laval, Quebec in May 1987 as a joint venture between University of Quebec's Armand Frappier Institute and MDS Nordion. The Centre is one of the world's leading facilities for research, training, operational demonstrations, and product and market trials of irradiation. A food irradiation research facility has also been established at the Agriculture Canada station at Ste.-Hyacinthe, Quebec.

In contrast to the rest of the world, there is no commercial food irradiation in Canada, in spite of considerable media attention to food poisoning. In 1998, for example, a food poisoning outbreak in Canada caused over 500 people to become ill. The tainted cheese containing the salmonella bacteria came from a single major food processing plant, from where it was shipped all across the country. This large, centralized plant is well suited to incorporate irradiation which, as for many medical products, could be done after the food is packaged.

Foods are normally approved for irradiation on an individual basis by Health Canada. Since 1984, only one food item has been approved: the mango — hardly a staple of Canadian diets. A petition to approve irradiation of poultry, originally submitted to Health Canada in May 1993, was still under review eight years later. No meats have been approved. As Health Canada dithers, Canadians continue to get sick and die from food poisoning. It is ironic that the country that is the leader in developing food irradiation technology is the slowest in accepting it.

INSPECTION AND GAUGING

Radiography

Just as X-rays are used to look at bones inside humans, radiography uses gamma rays to take "shadow" pictures of industrial components such as aircraft castings and welds in oil and gas pipelines to detect structural faults, impurities, and porosity. Radiography uses principally iridium-192 sources (half life of 74.2 days) which are portable, do not require any electrical power, and whose gamma rays are more penetrating than X-rays. More than 800 iridium sources are used in Canada each year, with each one performing 5 to 50 radiographs per day during a useful life of three to six months.

To inspect pipe welds, for example, a crank-out camera is used as shown in Figure 12-3. A shielded container houses a radioactive source, which can be propelled along a guide tube using a crank and long cable by an operator, who remains a safe distance from the container. A strip of film placed around the pipe provides a radiograph of the weld, revealing any defects. More than 140 companies in Canada are licensed to perform radiography. This helps to produce high-quality piping systems and reduces the likelihood of failures and potentially serious consequences such as environmental contamination or explosions.

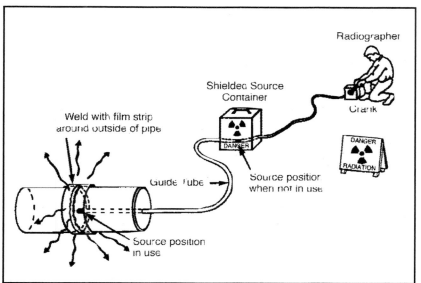

Figure 12-3: Crank-out radiography cameras of this type have many applications, such as checking the integrity of welds in oil and gas pipelines.

Nuclear Gauges

Because radiation can be detected and measured with great accuracy, radioisotopes make excellent tracers and gauges. There are two types of nuclear gauges: fixed and portable.

Fixed gauges are most often used in factories to monitor production processes and ensure quality control. A fixed gauge consists of a radioactive source housed within a shielded holder and placed at a critical point in the process. When the shutter in front of the source is opened, a beam of radiation is directed at the material being processed. A detector mounted opposite the source measures the radiation, which is diminished in proportion to the amount and density of material in between.

A significant advantage of nuclear gauges is that they do not need to be in contact with the material being controlled, and thus can be used on high-speed processes and on materials with extreme temperatures, pressures, or containing harmful chemicals.

Radioisotope gauges for measuring mass per unit area, called thickness gauges, are unequalled in their performance and are used in almost every kind of industry in which sheet material is produced (see Figure 12-4). In the paper industry, the accuracy and quick response

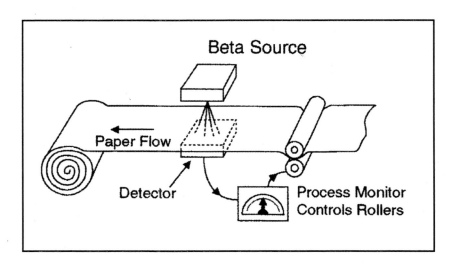

Figure 12-4: Schematic of a continuous-process paper-thickness gauge. Many other products such as textiles, sheet metal, and even cigarettes depend on high-speed, high-accuracy radioactive gauges for their production.

of these gauges is vital as the paper machines operate at high speed. Similarly, the production of steel plate in modern rolling mills could not be done without accurate measurement of thickness at every moment of the process and automatic control of the rolling stands. In the plastics industry, radioisotope gauges are used to improve the uniformity of the product, thus yielding savings in raw material and reducing costs for rejects. Isotopes commonly used for this purpose are cesium-137, krypton-85 and strontium-90.

Another application is the monitoring of material flow in a pipe or chute. In coal-fired electrical-generating stations, for example, the flow of coal into the furnace can be monitored by a radioactive gauge. If a blockage occurs an alarm sounds, allowing action to be taken to remove the blockage. Comparable methods, based on relating changes in the attenuation of radiation to changes in the materials it passes through, are widely applied to measure the levels in containers and tanks of all kinds.

Nuclear techniques make possible on-line determination of contaminants such as sulphur and nitrogen (the causes of acid rain) in coal. This method has become routine in the coal industry, with hundreds of millions of tonnes of coal analysed every year, preventing considerable contamination of the atmosphere.

In galvanizing or tin-coating of steel plate and cans, the exact amount of coating material must be applied as too much material is expensive, and undercoating results in early corrosion. Nuclear gauges allow the coating processes to be controlled to tight limits, and up to 10% material savings can be realized over other methods.

Portable gauges are used in many industries such as agriculture, construction, and civil engineering to measure moisture in soil and the density of asphalt or compacted clay. There are two basic methods, direct transmission and backscatter.

In direct transmission, which is more accurate, the source is placed in a tube and inserted beneath the surface through a hole created for the purpose. Radiation is transmitted to a detector in the base of the gauge, where it is measured, and the density of the soil calculated based on the amount of radiation detected.

The backscatter method is quicker as it eliminates the need for an access hole, but it is less accurate. Radiation is directed beneath the surface where some of it is reflected or scattered back to the detector in the gauge.

Each nuclear gauge uses one or two small radioactive sources such as cesium-137, americium-241, radium-226, or cobalt-60, placed in a special capsule which can be as small as the eraser on a pencil or as large as the tube inside a roll of paper towels. The housing that contains the radioactive source is constructed of heavy metal to provide shielding. As these sources contain relatively large amounts of radioactivity, which can be harmful, they must be managed carefully and disposed of properly.

Radioisotopes are also used as tracers to find lost buried pipes or leaks in pipes or storage tanks. A short-lived gamma-emitting radioisotope is flushed through the pipes with water while detectors are stationed on the ground above to trace the flow path and reveal any leaks.

OTHER APPLICATIONS

Agriculture

For centuries, humans have worked industriously to increase the quantity and quality of food crops. Nuclear technology has made a significant contribution to these efforts. Over the past 50 years, a number of plant breeding programs have been undertaken using mutation induction with radiation to increase disease resistance, improve yields, strengthen stems so the plants can better withstand storms, and increase winter hardiness. Some of these developments have had a significant impact, such as in Pakistan, where a new cotton strain was released in 1983. Cotton production in the country approximately doubled and the crop value of this new type of cotton was more than $2.3 billion in 1988 and 1989.

Water is one of the most important factors in crop production. Nuclear methods allow continuous monitoring of the moisture content of soil, which helps to determine the amount of water being consumed by plants and evaporation. In this way, soil scientists not only improve crop yields but, in some cases, up to 40 percent of the water can be saved.

Geology

Radioactive sources are routinely used in geophysical well logging to explore the earth's subsurface. Density, or gamma-gamma, gauges emit gamma radiation from cobalt-60 or cesium-137 sources and correlate the response to the density of the surrounding rock formation. Density, in turn, is a good indicator of porosity, a key parameter for finding rocks with potential for containing oil or gas. In a similar way, neutron gauges determine water content in rock formations because neutrons are absorbed by the hydrogen in water molecules. Radioisotopes are also used as tracers in tracking the connection between oil wells and the progress of secondary and tertiary oil-recovery processes.

Civil Engineering

Neutron and density gauges have been used for over 40 years to make civil engineering, agricultural, and hydrological measurements. Civil engineers, for example, use such gauges to establish how soils in road beds, under building and bridge footings, and in landfills have been compacted and how they will deform under loads.

Insect Control

Some insects have long been a nuisance and threat to human and animal health. Controlling insects with chemicals poses serious problems of environmental pollution, and generally is not very effective, as insects develop resistance to the insecticides and pesticides. An innovative method of controlling insects without having to use damaging chemicals is the sterile-insect technique. Large quantities of the particular insect are bred and then sterilized using gamma irradiation. When released into the native population, they mate with wild insects but no offspring are produced. This technique, which is the only environmentally sound method available, is most effective when the sterile insects can be produced in large numbers, and the native insect population is relatively small and isolated from other infestations. The first successful application of this method was to eradicate the screwworm, a devastating pest of domestic animals and wildlife, from the island of Curacao in 1954. Later, the screwworm was also eradicated from the USA and then Mexico. Texas ranchers estimated that the program saved them about $150 million annually.

Tsetse flies, which slowly destroy livestock by transmitting a parasitic disease, have prevented settlement of large areas of Africa. It is estimated that the human form of this disease, sleeping sickness, which eventually causes death, afflicts about 300,000 people in Africa. The tsetse fly has recently been eradicated from the once heavily infested main island of Zanzibar. Tsetse flies were bred in a special fly factory and the male flies were sterilized using low doses of gamma irradiation from cobalt-60 or cesium-137. Almost 8 million sterile flies were dropped by air over infested areas.

This method of eradicating problem insects was developed by entomologists, Edward Knipling and Raymond Bushland, who received the World Food Prize in 1992 for the achievement.

Safety

The smoke detector is so effective in saving lives that many municipalities make it a legal requirement to have them installed in new homes. The ionization-chamber smoke detector uses a radioactive material, generally americium-241, to ionize the air between two metal plates. An electric voltage is applied across the plates causing a small electric current. When smoke enters this space it causes the current to decrease, triggering an alarm.

Tritium (hydrogen-3) is used for lighting at runways, exit signs, and for situations where electrical light could constitute a fire hazard such as in coal mines or grain elevators. Tritium emits radiation which activates phosphors that produce light.

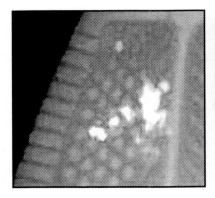 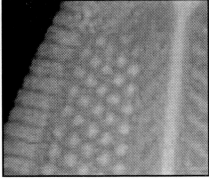

Nray Services, Pratt & Whitney Canada

Figure 12-5: Neutron radiographs of the same portion of two jet-engine turbine blades. The white traces near the top of the left blade indicate it is defective. The defective blade could overheat and fail causing the engine to fail.

DETECTING LAND MINES

Long after the wars in which land mines were used are over, they continue to kill and maim thousands of people every year. A big problem with detecting land mines is discriminating the mines from the many other objects in the ground. For example, tin cans have a similar magnetic signature as a metal land mine and rocks and stones of various types appear on ground penetrating X-rays. Mines made of plastic parts are particularly difficult to identify. Canada is not only a leader in international efforts to ban land mines but is also a leader in research aimed at finding efficient methods of detecting and removing them.

One research program is looking at directing neutron beams into the soil from a specially designed robotic vehicle. Neutrons encountering the high concentrations of nitrogen typical of explosives cause distinct gamma rays to be emitted, which can be detected. This technology is applicable to finding both metal and plastic mines.

Nuclear techniques also help to make airplanes safer. Today's jet engines operate at high temperatures and therefore, as shown in Figure 12-5, they are cast with built-in internal cooling channels. If any of these channels are blocked, the blade may fail due to overheating and cause the whole engine to fail. To ensure the blades are properly made, each one is inspected by neutron radiography using reactor neutrons. This is the only reliable method of detecting blockages in the cooling channels as X-rays are not effective.

Archaeology

Radioactive age dating has become a powerful scientific tool. Carbon-14 (half-life of 5,730 years) is constantly being produced by cosmic radiation striking the nitrogen in the atmosphere. Living plants take up the radioactive carbon as they absorb carbon dioxide. When the plant dies, the uptake ceases and the carbon-14 decays. The longer the plant is dead, the more carbon-14 will decay, and by comparing this amount to normal carbon, the elapsed time since the death of the plant can be calculated. This method has been widely used to date objects containing carbon which are from 1,000 to 40,000 years old, such as soils, shells, marine sediments, trees, archaeological artifacts, bones, and textiles. This method was used to determine the age of the famous Turin Shroud. Similar methods using other radioisotopes have been used to date rocks and unravel the history of the earth.

CASE HISTORY: THAILAND

Thailand has no nuclear reactors and no nuclear power program, however, considerable effort has been expended in applying radioisotopes and nuclear technology to improve their standard of living. To direct these efforts, Thailand established the Office of Atomic Energy for Peace in 1961, which currently has a staff of about 350. Some of their projects are described below.

A popular Thai delicacy is nham, fermented raw pork, which is sold in roadside stands. A health study in 1971 showed that 20% of the nham was contaminated with salmonella. Since then, trichinosis, a disease caused by eating undercooked pork, has also become a concern. A subsequent research program showed that the risk of salmonella could be significantly reduced by irradiation, with no change in taste, texture, or appearance. Irradiated nham is now available and has proven popular with Thais, who are willing to pay the slight extra cost to protect their health.

Irradiation of food and other agriproducts has considerable potential in Thailand because of its hot climate and strong agricultural base, particularly for high-cost export products such as mangoes, shrimp, and cut flowers. Canada helped Thailand construct a food irradiation research and demonstration centre, which became operational in 1989. MDS Nordion designed the cobalt-60 facility, managed the project, and trained the staff. The centre has helped Thailand improve its agricultural exports and also serves as a research centre for southeast Asia.

Other nuclear-based methods are being used to modernize Thailand's agricultural sector. It was found that the birth rates of water buffalo, which play an important role in cultivating crops in traditional farming, were stagnating. Based on results from radioimmunoassay testing of progesterone, methods have been developed that have improved birth rates in individual cows by 50 to 300%.

Fruit flies were having a devastating effect on exotic fruit growing in the northeast, with up to 90% of some crops being infected. In the mid-1980s, a fruit fly sterilization program was initiated, in which fly pupae were irradiated and then sterilized fruit flies were released into the orchards. The program was successful and was part of the solution to stopping the black-market growth of poppies for the heroin trade.

absorbed by crops in different soil types. Scientists are using these findings to select the plant varieties that are best suited to specific soil and moisture conditions. Rice, in particular, has received much investigation.

In another project, special neutron probes were inserted into the soil to determine the water content at various depths. The information was used to determine the soil's water storage capacity so that exactly the right amount of irrigation water would be applied and at the optimal rates. Similar probes measure soil density to determine how effectively plant roots penetrate the soil. Isotopes are introduced into the biological system and their subsequent movement is measured by instruments (similar to medical diagnostic methods). Such studies determine how fertilizers and other nutrients are technology as applied to industry, agriculture, and natural resources impacts our daily lives. There are many other examples and more applications are being discovered all the time.

Chapter Thirteen

Uranium: the Nuclear Fuel

The single most dominant image of nuclear technology is the reactor. It is in the bowels of this structure where the nuclear reaction takes place, and like a throbbing heart, it sends electricity pulsing along a network of transmission lines. Reactors are the most visible part of nuclear power, but there is much more.

To gain a complete understanding of nuclear power it is necessary to look at the fuel that drives it and the entire process, or life cycle, through which uranium passes, as depicted in Figure 13-1. In Canada, the **nuclear fuel cycle** begins with the mining of uranium ore from the ground. The ore goes through a complex milling, refining, and conversion process and then it is manufactured into nuclear fuel. The fuel is loaded into CANDU reactors where nuclear reactions generate electricity. After the fuel is consumed, it is removed from the reactor and stored on-site for a number of years while its radioactivity and heat subside. The feasibility of centralized long-term spent fuel storage will also be assessed. At some point in the future the spent fuel will be encapsulated in sturdy, leach-resistant containers and placed in permanent disposal by burying it deep underground from where it originated, thus completing the cycle. In addition, the transportation of nuclear materials is an important consideration. It is essential that all of these steps be conducted in a manner that does not endanger the public or the environment.

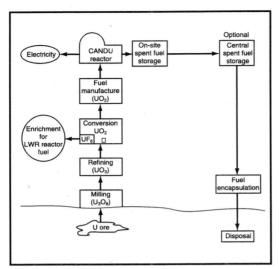

Figure 13-1: The Canadian uranium life cycle from mining of uranium ore through preparation of fuel to creation of electricity in CANDU reactors to final disposal deep underground.

The nuclear fuel cycle for light-water reactors is different from the Canadian approach as it includes enrichment of the uranium-235 in the fuel. The US fuel cycle also differs from the Canadian cycle in that long-term, central spent-fuel storage prior to disposal is an integral part of their waste management program. The French and British fuel cycles differ again by including reprocessing of their spent fuel. This involves the removal of some fissile elements from the spent fuel so they can be incorporated into fresh reactor fuel. They do not plan to have long-term central spent-fuel storage.

Some of the steps comprising the nuclear fuel cycle are described in earlier chapters. Reactors and how they operate are described in Chapters 6 and 7; waste management, including interim on-site storage, long-term central storage, and permanent disposal, is described in Chapter 10. The remaining components of the nuclear fuel cycle, namely, mining, refining/conversion, fuel manufacture, reprocessing, and transportation are described in this chapter.

URANIUM MINING

Uranium is a fascinating element. It was discovered in 1789 by Martin Klaproth, who named it after the recently discovered planet Uranus. Interestingly, Klaproth also discovered zirconium, another important element in nuclear technology. Uranium is one of the more common heavy elements in nature, 500 times more abundant than gold and about twice as common as tin. It is present in virtually all rocks and soils as well as in rivers and oceans. Traces are found in food and human tissue. Granite, which makes up about 60% of the earth's crust, averages about four parts per million (ppm) uranium, although the concentration is highly variable. Phosphate rock used to produce fertilizer can contain as much as 400 ppm uranium, and some coal deposits contain up to 1,000 ppm. Long used to add colour to glass, ceramics, and porcelain dentures, it is only in the past half century that it has become a valuable energy source. Uranium deposits with concentrations of about 1,000 ppm and greater of uranium may be considered "ore", that is, they may be economic to mine.

Gilbert Labine (1890-1977): was born in Westmeath, Ontario, and attended the Provincial School of Mines at Haileybury, Ontario. He became a prospector and mine developer founding Eldorado Gold Mines. In 1930, he discovered Canada's first major uranium deposit at Port Radium on Great Bear Lake in the Northwest Territories (where this picture was taken in 1935). Eldorado was taken over by the government of Canada in 1944, and Labine turned to other successful mining ventures. At the time of his death he was a respected and wealthy senior statesman in the mining community.

Cameco

Prospecting, the search for valuable minerals, holds a romantic place in the history of Canada. Early discoveries of uranium were made in the traditional method by grizzled prospectors braving the wilderness on foot and canoe. But modern times have led to modern methods. Today, airplanes specially equipped with radiation detectors fly low-level surveys over areas that, based on geological maps, appear promising. Once an area has been identified as having potential, field crews investigate it on the ground using hand-held radiation detectors such as Geiger-Muller counters (see Appendix A). Water, soil, and vegetation samples are also collected and analyzed for their uranium content.

Considerable detective work is required to find uranium deposits.

181

For example, if the ore is close to the surface, geologists reconstruct the movement of past glaciers to determine from where uranium bearing boulders may have come. If uranium deposits are hidden deep underground, geophysical methods such as electrical-resistivity surveys are conducted to find materials like graphite, which are good conductors of electricity and are often associated with uranium deposits. Once the search has been narrowed to a small target area, drill rigs are brought in, usually by float-planes, and bring up samples of rock from the subsurface. Chemical analyses of the samples confirm whether uranium is present in economic concentrations and help define the size and shape of the ore body.

Canada has been a major world producer of uranium since the global demand for this material developed. Today, the main producing area is northern Saskatchewan, although other areas have been active in the past. Canada is the world's leading exporter of uranium and hosts three of the top ten producing mines in the world. To place this into perspective, Canada's production of 10,922 tonnes of uranium in 1998 contained more than twice the energy available from Canada's total annual oil production.

The story of Canada's uranium mining industry began in 1931 in the cold and forbidding north, when prospector Gilbert Labine discovered pitchblende, a uranium-bearing mineral, near the shores of Great Bear Lake in the Northwest Territories. His discovery resulted in the development of a mine at Port Radium, North West Territories, in 1932, and a refinery in Port Hope, Ontario, in 1933. These facilities were owned and operated by Eldorado Gold Mining Company. The original objective was to produce the rare and precious element, radium, which is found in uranium ore, and was felt to be a miracle cure for cancer. It commanded prices as high as $75,000 per ounce until the bottom fell out of the market in the late 1930s.

During World War II, the demand for uranium took centre stage in order to supply the American nuclear weapons development program. The federal government took over Eldorado and formed a crown corporation, which was later renamed Eldorado Nuclear Limited. The war-time demand for uranium continued in the post-war era, but driven by the appetite of electrical power reactors. Port Radium produced uranium until the mine was closed in 1960.

Ontario has also seen considerable uranium activity. The Bancroft area witnessed radium mining in the 1920s and 1930s, and lived through two uranium booms from 1956 to 1964 and from 1976 to 1982. Faraday, Bicroft, and Madawaska Mines, all abandoned today, produced about 6,700 tonnes of uranium oxide (U_3O_8) using underground mining methods.

Elliot Lake, Ontario, became an uranium boom town virtually overnight. In 1954, Denison Mining intersected uranium in exploration drilling, and a brief three years later the first mine was in production. From there, Elliot Lake grew rapidly and soon gained a reputation as the "Uranium Capital" of the world. Two major companies, Denison Mines Ltd. and Rio Algom Ltd., operated five mines (Denison, Stanrock, Panel, Quirke, Stanleigh) and their accompanying mills. The ore, with a grade of approximately 0.1 to 0.2% uranium oxide, was mined from a depth of about 170 to 950 metres using underground mining methods.

As a result of the continuing foreign military demand for uranium, Canada's uranium mining industry continued to grow in the post-war years until 1959, when more than 12,000 tonnes of uranium oxide were produced, yielding $330 million in export revenue, more than any other mineral. Over the next few years, however, the military demand declined and the number of mines operating in Canada decreased to four. Uranium exploration waned in the 1960s and the Canadian government conducted a uranium stockpiling program until 1974 to support the industry. Thereafter, the uranium industry again experienced growth due to the increasing demand for electricity-generating nuclear reactors.

After three decades as the uranium capital of the world, Elliot Lake bowed to the inevitable fate of all mining centres. Unable to withstand the strong competition from the much higher-grade ore bodies in Saskatchewan and their lower production costs, the Ontario mines were decommissioned in the early to mid 1990s, having produced over 550,000 tonnes of uranium oxide. In 1996, with the closure of Stanleigh Mine, Saskatchewan became the sole province producing uranium.

Northern Saskatchewan
The Athabasca Basin of northern Saskatchewan has been the site of all major Canadian uranium discoveries in the past 20 years. The first northern Saskatchewan uranium deposits were discovered in the early 1950s and Eldorado began mining at Beaverlodge Mine in 1953.

Cameco

Figure 13-2: The McArthur River mine. The ore is pumped up from the mine mixed with water and transported to the Key Lake mill in the tanks shown on the truck.

In 1968, the Rabbit Lake deposit was discovered in northern Saskatchewan by Gulf Minerals Ltd. and the German-owned Uranerz Exploration and Mining Limited, and by 1975 the mine was in operation. Rabbit Lake was subsequently sold to Eldorado Nuclear in 1981.

In 1975, the French-owned company Amok Ltd. discovered the Cluff Lake deposit and mining commenced in 1980. 1975 also marked the discovery of the large Key Lake deposit by Uranerz Canada Ltd. A 50% interest in this deposit was sold to Saskatchewan Mining Development Corporation, a provincial crown corporation, and production began in 1983. During the latter half of the 1990s, these three Saskatchewan mines were the only active uranium mines in Canada.

Mining in the 1980s and 1990s was primarily by the open-pit method as the deposits were near the surface. Surface mining is more economical than underground mining and, combined with the very high ore grade found in the Athabasca Basin (about 2 to 10%), made this uranium very competitive in world markets. The high ore concentrations also require that great care be taken to ensure radiation protection for workers.

184

In 1988, the federal and Saskatchewan governments agreed to the amalgamation of their respective crown corporations, Eldorado Nuclear Limited and Saskatchewan Mining Development Corporation. The resulting company, Cameco Corporation, is currently the world's largest uranium producer, having recently purchased most of Uranerz's Canadian holdings. The other major player in Saskatchewan uranium mining is Cogema Resources Ltd., a French-owned company which purchased all of Amok's Canadian interests as well as some of the Uranerz holdings. Cogema also produces uranium in other countries and ranks a close second to Cameco in total uranium production.

Canada's uranium production in 2000 was 10,693 tonnes which represented about 30% of world output. Almost half of this production came from the McArthur River mine (Figure 13-2) alone. Canada's uranium ore reserves are about 15% of the world total.

Key Lake has been the highest grade and largest uranium mine in the world, but its useful life has come to an end and mining has ceased. Rabbit Lake Mine is also coming to the end of its reserves. New ore bodies, however, have been found and four new mines will be in production over the next few years (see Table 13-1). The McClean Lake mine commenced production in July 1999.

Table 13-1	SUMMARY OF SASKATCHEWAN URANIUM MINES			
Mine	Reserves *	Grade**	Owner	Dates of Operation
Beaverlodge			Eldorado	1953 - 1982
Rabbit Lake	37	1.7	Cameco	1975 -
Cluff Lake	20		Cogema	1980 -
Key Lake	183	2	Cameco	1983 - 97
McClean Lake	23	3	Cogema	1999 -
Cigar Lake	160	14	Cameco	2000 -
McArthur River	845	21	Cameco, Cogema	1999 -
Midwest Lake	16	4	Cogema	2003 -

* 1,000 tonnes ore ** % U_2O_3

As with the existing three mines, it will be mined primarily by open pit methods. Production in the other new mines will be by underground mining.

Cigar Lake was discovered in 1981 and construction started in the summer of 1997. The Cigar Lake ore body is the richest in the world, with local ore grades reaching more than 25% U_3O_8, and an average grade of over 15%. It is also one of the largest, with geological reserves totalling 136,000 tonnes of U_3O_8. It will be mined from tunnels above and below the ore zone, using a raise-boring technique on ore that is frozen before boring commences. The ore will be crushed, ground, mixed with water, and then pumped as a slurry to the surface for transportation to the mill. Special remote-control mining methods will be necessary due to the high radiation fields of the rich ore. Currently, a test mine is in operation. About half of the Cigar Lake ore will be milled at McClean Lake, and the other half will be shipped to Rabbit Lake for milling.

In 1988, the massive McArthur River ore body was discovered. With geological reserves totalling 845,000 tonnes of U_3O_8 at an average grade of close to 20% U_3O_8, this is the largest known uranium deposit in the world. At a depth of 550 metres, it is mined by underground methods similar to those at Cigar Lake; it uses the Key Lake mill. The mill tailings, or wastes, are placed into the former Deilmann pit of the Key Lake mine, which has been converted into a tailings management facility using the "pervious surround" method, which has been successfully used at the Rabbit Lake mine since 1985. Mining at McArthur River began in 1999; in 2000, it produced 5,000 tonnes of uranium.

With their enormous reserves and exceptionally high uranium concentrations, about 100 times the world average grade, the Cigar Lake and McArthur River mines are the future of Saskatchewan uranium mining and eclipse any other deposits in the world.

Another much smaller mine, Midwest, is currently obtaining approvals, and is expected to go into production in 2003. It will also use the McClean Lake mill.

The enormous bounty of uranium buried in the Athabasca Basin is almost beyond reckoning and can provide substantial wealth to the province of Saskatchewan and Canada for decades to come. It is estimated that, at 1997 prices, there is $25 billion of uranium in known deposits.

International
Virtually all of the western world's uranium is supplied by eight countries: Canada, Niger, Australia, Namibia, South Africa, United States, France, and Gabon. Since 1988, the former Soviet Union and

China have also supplied significant quantities of uranium to western world markets.

In Europe and Japan, limited quantities of reactor fuel are obtained by reprocessing spent fuel. It is estimated that reprocessing will supply less than 5% of the western world's uranium requirements in the next decade.

In 1994, the US and Russia agreed to convert highly-enriched uranium from dismantled nuclear weapons into low-enriched uranium for use in nuclear power plants. In 1997, Russia was scheduled to deliver low-enriched fuel containing about 6,400 tonnes of U_3O_8, this was to increase to about 8,600 tonnes in 1998, and to about 10,900 tonnes by 2005.

Uranium demand is easy to predict because it depends solely on the number and size of nuclear reactors, information which is well known. The annual consumption by the western world in 1997 was 67,000 tonnes which is expected to grow to 76,000 tonnes by 2007. The western-world electric utilities accumulated large inventories of uranium during the late 1970s and 1980s, but these inventories were depleted by 2000. Even with uranium from the dismantlement of nuclear weapons, the production from existing mines will not be sufficient to meet demand. Without the new mines that are being developed in Canada and elsewhere, there would be a shortfall in the next decade. Mineable reserves are in place to provide many decades of uranium fuel.

Health and Safety

The greatest risk associated with uranium mining is the same as is faced in all mining operations, namely, industrial accidents. Safety in uranium mines is further complicated by the presence of the radioactive gas radon and its daughter products, which can irradiate lung tissue when inhaled. This is particularly a concern in underground mining because of the confined spaces. The potential hazard can be minimized by using powerful ventilation systems in underground mine spaces, special respirators in some cases, and limiting work time in some areas. All workers carry radiation monitors that are tracked to ensure radiation doses received by workers are well below the limits set by the regulatory agency, the Canadian Nuclear Safety Commission.

Export Policies

Canada has several policies which govern the export of uranium and nuclear technology. Administered by the Canadian Nuclear Safety Commission, these are designed to increase the benefits to Canada of developing its uranium resources and to prevent these resources and technologies from being used to manufacture nuclear weapons.

Federal guidelines state that at least a 30-year reserve of uranium supply must be available for all operating reactors in Canada, as well as for those planned or committed for the next ten years. In addition, sufficient uranium production capacity must be available for the domestic nuclear program to reach its full potential. Only after these requirements are met, may uranium be exported.

Federal guidelines also stipulate that uranium for export must be processed to the most advanced form possible, provided that such processing capability is available and is economically competitive with that available elsewhere. This guideline ensures that jobs related to more advanced processing, not just mining, stay in Canada.

URANIUM PROCESSING

Once the uranium ore is extracted from the ground, it must undergo a number of chemical processes before it is ready to be made into fuel for nuclear reactors.

Milling

Uranium ore is first treated in a mill at or near the mine site where the ore is ground to a very small size. Uranium oxide (U_3O_8) is then leached from the rock, usually using sulphuric acid, and dissolved. The solution is extracted and dried, or calcined, to yield a yellow powder, called yellowcake, which contains about 70-82% uranium by weight. The yellowcake, which represents only a small fraction of the mined ore, is then shipped to a refinery. The Key Lake mill is shown in Figure 13-3.

As is the case for all mining operations, large amounts of waste rock called **tailings** are created during the mining and milling process. These are a mixture of solids and liquids called a slurry. As the slurry is acidic due to chemicals used in the milling process, lime or lime-stone is added in the mill to neutralize it. The slurry is then discharged into a tailings management area, effectively a large pond, where solids precipitate out and water is drained off. The solid tailings that remain are similar in composition and hazard to the ore that was

Cameco

Figure 13-3: The uranium mill at Key Lake where uranium ore is crushed and uranium is extracted in the form of yellowcake for shipment to the refinery.

originally mined. They are, however, more mobile and must be properly contained.

Tailings are contaminated with radium and other hazardous materials. In 1996, there were about 120 million cubic metres of uranium tailings in Ontario and about 14 million cubic metres in Saskatchewan, with the latter amount steadily growing as mining continues. Uranium mines account for about 2% of all mine tailings produced in Canada.

At most uranium mines, dams are used in conjunction with natural topography to contain tailings within a management area. At Elliot Lake, methods have been developed for impounding the mine tailings and then treating the runoff water in a series of ponds where barium chloride is added to precipitate the radium still left in the water. After treatment, the water is released into the natural waterways of the area.

Similar methods are also being used in Saskatchewan. At the Key Lake mill, a tailings containment area lined with bentonite, a type of clay, has been created. The Key Lake tailings pond is shown in Figure 13-4.

189

Cameco

Figure 13-4: The Key Lake tailings pond is a modern facility built from a previously mined open pit.

At Rabbit Lake, the tailings are placed in a mined-out open pit mine using an innovative disposal method. Instead of using traditional impermeable liners like bentonite or plastics, the pit is lined with separate layers of sand and gravel, which are highly permeable to the flow of groundwater. The method takes advantage of the relatively permeable nature of the blast-fractured pit walls compared to the surrounding rock. De-watered tailings are placed into the pit and compacted. Seepage from compaction of the tailings is pumped from tunnels below the mine into the mill for reuse. The compacted, more dense, and less permeable tailings are surrounded by a less dense, more permeable envelope, through which groundwater will flow, thus bypassing the waste tailings and leaving them intact. A lake will be allowed to form over the top of the tailings area, eliminating the risk of airborne radioactive dust, reducing radon emissions, and avoiding inadvertent human intrusion. This "permeable-surround" or "hydraulic-bypass" concept will also be used for disposal of wastes from the McArthur River mine.

In earlier mining operations such as at Bancroft, however, tailings were generally not disposed of in as rigorous a manner and have left some environmental problems that persist to this day.

Refining

Yellowcake must be chemically refined to separate the uranium from the remaining impurities and convert it to uranium trioxide. This process was originally done at Port Hope, Ontario, but since 1983 has taken place at Cameco's facilities at Blind River, Ontario, the only uranium refining facility in Canada. Refining is a complex chemical process involving the following steps:

1- dissolution of yellowcake in nitric acid resulting in a uranyl nitrate solution
2- impurities are removed from the uranyl nitrate solution
3- water is evaporated from the solution
4- the uranyl nitrate is heated to form uranium trioxide powder

Care is taken to recycle chemicals at every step to ensure extraction of virtually all of the uranium and to reduce wastes. The nitrogen oxides produced in step 4, for example, are converted to nitric acid, which is then used again in step 1.

Finally, the pure uranium trioxide powder is shipped to Port Hope for the next processing step, called conversion.

Cameco

Figure 13-5: Aerial view of Cameco's uranium refinery at Port Hope, Ontario. The plant performs the 'conversion' step to produce uranium dioxide for CANDU reactor fuel and uranium hexafluoride to feed the uranium enrichment facilities in other countries.

Conversion

Cameco operates Canada's only **conversion** plant at Port Hope, Ontario (Figure 13-5), where two main products, uranium dioxide (UO_2) and uranium hexafluoride (UF_6), are produced. About 20% of the uranium received at Port Hope is made into UO_2 and is shipped to manufacturers who make the fuel pellets and bundles used in CANDU reactors. The remaining 80% is made into UF_6 for subsequent enrichment outside Canada for use in light-water reactors.

Ceramic-grade UO_2 powder is produced by a complex chemical process that starts by dissolving UO_3 from the refinery in nitric acid. The final UO_2, a brown powder, is shipped to Canadian manufacturers who make the fuel bundles used in CANDU reactors. The only waste generated in the conversion to UO_2 is liquid ammonium nitrate, which is concentrated and sold as a fertilizer to local farmers. This "nuclear" fertilizer contains less uranium and radium than many commercial fertilizers.

To produce UF_6, the UO_3 from the refinery is pulverized to a fine powder and then reduced with hydrogen into UO_2. In the next step, the UO_2 is mixed with hydrofluoric acid (HF) to make uranium tetrafluoride (UF_4). In the final step, the UF_4 is burned with fluorine gas to produce gaseous UF_6, which condenses to a solid white crystalline form as it cools. It is subsequently heated to a liquid for transfer into steel shipping cylinders where it is allowed to cool into a solid form, prior to shipment to enrichment plants in Japan, Europe, and the United States.

All emissions from the UF_6 plant pass through an acid recovery system and then a scrubber to capture and reduce fluoride, which is harmful to vegetation. The scrubbing solutions are treated to remove uranium and then lime is added to create calcium fluoride, the only waste from the UF_6 process. In past years this waste was discarded in landfills, but recently this material has been recycled to steel mills, where the calcium fluoride is used as a flux to separate impurities from steel.

In summary, wastes are produced in the refining and conversion operations carried out at Blind River and Port Hope. These wastes consist primarily of dilute sulphuric acid slurry, calcium fluoride sludge, ammonium nitrate, and miscellaneous garbage and scrap metal. Cameco Corporation has been proactive in reducing the amount of waste through a program of recycling and minimization.

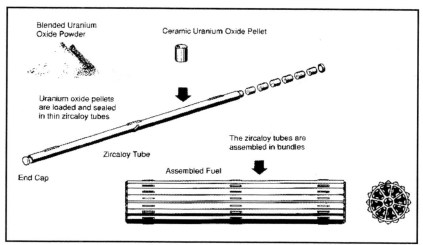

Figure 13-6: Schematic illustrating the steps in CANDU fuel fabrication. Uranium oxide powder is compressed to form ceramic pellets about 2 cm in length and 1 cm in diameter. The pellets are sealed into zircaloy tubes which are then welded together to form a fuel bundle.

FUEL FABRICATION

Zircatec Precision Industries Inc. at Port Hope, Ontario, and General Electric Canada at Toronto and Peterborough, Ontario, manufacture fuel bundles for CANDU reactors (see Figure 13-6). The process starts by sintering uranium dioxide in a hydrogen atmosphere at about 1800 °C into ceramic pellets about one centimetre in diameter and about two centimetres long. Thirty pellets are assembled into a long tube of zircaloy metal, known as a fuel element. Elements are assembled into fuel bundles which are about 50 cm long, about 10 cm in diameter, and weigh about 22.5 kilograms. The fuel bundles are designed to allow the flow of water through and around them. The fuel contains few impurities which could absorb vital neutrons. Zircaloy metal, an alloy of zirconium, was selected to contain the fuel pellets because it is highly corrosion resistant and absorbs very few neutrons compared to stainless steel. CANDU fuel bundles are not all exactly the same, although they are the same length.

URANIUM ENRICHMENT

Approximately eighty per cent of the uranium received at Port Hope is made into uranium hexafluoride and shipped to enrichment plants outside Canada. Using a process called gaseous diffusion, the concentration of uranium-235 is enriched to about 2 to 5% as required in light-water reactors. The USA, France, Britain and Russia operate gaseous diffusion enrichment plants; no enrichment is conducted in Canada. Uranium hexafluoride is heated until it is a gas and then is pumped repeatedly through porous metal filters with holes one-millionth of a centimetre in diameter. Uranium-235, which is very slightly lighter than uranium-238 because of its three fewer neutrons, passes through the filters marginally faster than the uranium-238. Thus, the gas on one side of a filter becomes enriched. By passing the gas through thousands of filters the concentration of uranium-235 is continuously enriched. Once complete, the enriched UF_6 is shipped to fuel manufacturers who convert it back to UO_2 and process it into fuel pellets and fuel bundles of the shape required for light water reactors.

Although the gaseous diffusion process consumes large amounts of energy, the enriched uranium still generates about 30 times more energy than is required to produce it.

A new enrichment technology, the gas centrifuge, has been developed that consumes only about 4% of the energy of an equivalent gas diffusion plant. Centrifuge enrichment plants are now operating in Britain, the Netherlands, and the USA. The UF_6 is heated until it is in gaseous form; it is then spun in a centrifuge at very high speeds. The heavier uranium-238 atoms move to the outside of the centrifuge, leaving the central parts slightly enriched in the lighter uranium-235 atoms. By repeating this process many times the uranium-235 concentration is increased to the desired level.

While still in the experimental stage, laser enrichment appears to be a promising technology. Since certain molecules containing uranium-235 and uranium-238 absorb different wavelengths of light, different frequencies of laser light can be used to excite and separate the two isotopes.

REPROCESSING

Reprocessing is to nuclear waste what recycling is to regular garbage: it removes some useable (fissile) materials from spent fuel that can be reused to generate more energy. This extends fuel supplies and reduces the radioactivity and quantity of waste to be disposed.

As fuel is burned in a CANDU reactor, some, but not all, of the uranium-235 is depleted. At the same time, some plutonium-239, a fissile material, is created. Thus, the amount of fissile material in irradiated fuel is only decreased by about 30% from that in the fresh fuel. Irradiated fuel is removed from the reactor, not because there is no more energy left to extract, but because too many neutron-absorbing fission products have built up (see Table 10-1).

Instead of mining more uranium from the ground, the plutonium and uranium-235 could be extracted, that is, reprocessed, from the spent fuel and reused to make new fuel and generate more power.

Canada has never undertaken commercial-scale reprocessing and has no plans to do so in the foreseeable future because of the high costs involved as well as the concern that plutonium could be diverted and manufactured into nuclear weapons. In the USA, no reprocessing of power-reactor fuel is currently being undertaken, although it has been done since the 1940s as part of their weapons program.

Reprocessing is a complex chemical procedure that involves chopping up spent-fuel elements and dissolving the pieces in nitric acid. Chemical separation of uranium and plutonium then follows. These steps must be done by remote control in shielded rooms due to the high radioactivity of the spent fuel. The extracted uranium is returned to the conversion step prior to fuel fabrication; the plutonium goes straight to the fuel fabrication plant.

Commercial reprocessing forms an integral part of the nuclear fuel cycles of some countries like England and France. There are eight reprocessing plants worldwide with a total of over 5,000 tonnes per year capacity. In the past 40 years, more than 55,000 tonnes of spent fuel from electrical power reactors have been reprocessed. Great Britain has two plants: an older 1,500 tonnes/year plant has been in operation for over 40 years at Sellafield, primarily for metal fuel elements; a newer thermal oxide reprocessing plant (THORP) was commissioned in 1994. In France a 400 tonne/year plant reprocesses metal fuels at Marcoule. At La Hague, oxide fuels have been reprocessed since 1976 by two 800 tonne/year plants. In addition, India has a 150 tonne/year plant, and Russia has a 400 tonne/year plant. Japan operates a small 100 tonne/year plant and a much larger one is nearing completion. Countries like Japan are turning to reprocessing because they lack indigenous fuel sources and they wish to be energy independent.

195

TRANSPORTATION

Movement of radioactive materials is commonplace, due not only to the many shipments involved in the nuclear fuel cycle, but also to the thousands of radionuclides used in medicine and industry. In Canada, the Canadian Nuclear Safety Commission licences over 4,000 users of radioactive materials and there are about one million shipments of nuclear materials each year. These shipments are made in accordance with performance standards that are based on international standards developed by the IAEA. These standards are very rigorous and are designed to protect the public even in the case of severe accidents. Five different package types can be used, depending on the type of material to be transported and its level of radioactivity. They are described below in ascending order of radioactive content.

1. Excepted packages (also called limited activity packages) are for radioactive material with a very low activity and a very low hazard. Even if released in an accident, the amount of material does not pose a significant hazard. These packages do not require labels or special markings and could be, for example, an industrial-strength cardboard box.

2. Industrial packages contain what are called low-specific-activity materials or surface-contaminated objects such as uranium concentrates and related compounds that are shipped in large quantities in Canada. The hazards, even if materials are released in an accident, are minimal, but there is a potential for nuisance, economic disruption, or environmental degradation. A typical industrial package might be a steel drum marked with a Class 7 "Radioactive" placard.

3. Type A packages contain medium activities of radioactive materials such as radioisotopes being shipped to hospitals or radioactive sources used in research or in industrial applications. The hazard of an accident is controlled by the amount of radioactivity and the strength of the package. A Type A package may be a metal, plywood, or cardboard box or drum with internal foam padding, lead shielding, absorbent material (if a liquid), and other containment features. The package must bear labels and the transport vehicle must be placarded. Ontario Power Generation uses a Type A Container to ship low-level waste. To obtain a CNSC licence, scale modelling and/or computer modelling must demonstrate that it can withstand a fall of 1.2 metres onto a hard surface and an attempted puncture by a steel bar. If the container is to be used for gas or liquid waste, it must undergo more

DUAL PURPOSE CONTAINER

An interesting development is the use of containers for more than one purpose. A German company has developed a new steel and concrete cask called CASTOR for the dual purpose of transporting and storing spent nuclear fuel. As part of the licensing process required for type B package, they were subjected to two different drop tests. The first test consisted of a 9-m drop of the cask on its sidewall onto an unyielding foundation. No deformations of the cask body occurred. The second test was a 1-metre drop onto a steel pin without shock absorbers. The observed minor deformation on the outer surface was in agreement with predicted computer calculations and did not impair the integrity of the cask.

rigorous testing including a more forceful puncture test and a fall from a nine-metre height.

4. Type B packages contain large activities of radioactive material and must be licensed by the Canadian Nuclear Safety Commission. Typical materials include cobalt therapy sources, radiography cameras, and irradiated nuclear fuel. The hazard of an accident is controlled by the strength and robustness of the packages, which are constructed of steel with depleted uranium or lead as shielding and weigh from 25 kg to 35 tonnes. Type B packages must be labelled and the transport vehicles must be placarded.

Ontario Power Generation has designed several Type B packages, including a two-bundle cask for irradiated fuel, a bulk resin flask for ion-exchange resins and higher level waste, and a tritiated heavy-water transportation flask. All Type B containers must be able to withstand a nine-metre fall onto a hard surface, a one-metre drop onto a steel spike, 30 minutes in a 800 $^{\circ}$C fire, and eight hours immersed below 15 metres of water.

Sometimes the radioactive material to be shipped in Type A or B packages is placed into a special capsule or is put into a ceramic-like form to provide additional containment and safety.

5. When fissile materials are shipped, the word "Fissile" or the letter "F" is added to the description, as in a Type AF package.

The labels placed on these packages must indicate the maximum radiation level at the surface of the package and at one metre distance. Transportation regulations and practices in most countries are similar as they are based on IAEA guidelines.

Chapter Fourteen

The Fission Future

Nuclear energy today is at a cross roads. On the one hand, it can be argued that the use of nuclear power should increase because it is reliable and environmentally clean and produces no global-warming emissions. But in many countries there is opposition to nuclear power. In Sweden and Germany, governments have mandated that their nuclear power reactors be phased out. On the other hand, many countries, particularly, those on the Pacific Rim such as Korea, Japan, and China, are vigorously expanding their nuclear programs. And although no new reactors have been built in the United States in the past decade, the use of nuclear power is at an all-time high due to increasing capacity factors.

So where is nuclear headed? Is nuclear power on the verge of a renaissance, or will it gradually fade away over the next half century?

A complex combination of social, political, technical, economic, and emotional factors will determine the future of nuclear power. In making such decisions we need to look beyond the status of nuclear power as it stands at the present moment and also consider how it might evolve in the future.

Nuclear power, more than any other power source, involves high technology, an area where ideas are always popping up, new theories being developed, and innovative spin-off technologies being discovered. Thus, reactor design is constantly evolving and improving. Let us look at what advances are being pursued that might have an impact on decisions regarding the future of nuclear power.

ADVANCED CANDU FUEL CYCLES

Because of its excellent neutron efficiency, the CANDU reactor can operate with low concentrations of fissile material, allowing considerable flexibility in developing advanced fuel cycles. Some areas which are currently receiving consideration for further research and development include the following:

Slightly Enriched Uranium
Uranium with slight enrichment (between 0.8 and 1.2%) can be used in CANDU reactors without changes in the reactor design. This can increase the energy produced per kilogram of fuel by a factor of two to three while reducing fuel costs by about 30%.

Direct Use of PWR fuel In CANDU (DUPIC)
This fuel cycle is one in which spent fuel from light-water reactors would be used in a CANDU. This is possible because once spent fuel is finished in PWR and BWR reactors, it still contains about 1.5% fissile material which, although too low for use in LWRs, is suitable for the CANDU. First, neutron-absorbing fission products would need to be removed from the spent fuel. This would be achieved by removing the cladding from the fuel in a hot cell and then treating the fuel by a dry process. The resulting powder would be pressed into pellets and then made into new fuel bundles. This process is attractive because there is no chemical separation of plutonium, the material from which a nuclear bomb can be made. The resulting fuel would yield about twice the energy as was produced by the original fuel.

The DUPIC concept is currently being studied by Korean utilities, who would like to optimize their electrical production by first burning fuel in PWR stations and then reusing it in their CANDU stations. Thus, CANDU reactors can be viewed as being complementary to light-water reactors, rather than as competitors. This two-stage fuel cycle would triple the amount of energy that can be extracted from the original uranium.

Thorium fuel cycle
This been the subject of research in Canada for many years. Thorium-

232, although itself not fissile, can transform by neutron absorption into uranium-233, a fissile element. In nuclear parlance thorium-232 is **fertile**. In a thorium-fuel reactor, some fissile material (natural uranium or slightly enriched uranium) is required initially to make the reactor critical. Once the system is started, significant quantities of uranium-233 are produced and, in turn, fission. The production of uranium-233 is more efficient than the equivalent production of plutonium-239 in the natural-uranium fuel cycle, and under optimum conditions can create almost one atom of uranium-233 for each atom of uranium-235 destroyed. This is possible in a CANDU reactor but not in a light-water reactor, because of its excellent neutron economy and could make it a near breeder. That is, the CANDU could create almost as much fuel as it consumes, a remarkable accomplishment that fossil-fuel power plants can not match.

Two thorium fuel cycles are possible: with and without reprocessing. The former yields much better fuel utilization but is more complex and may be more costly on a unit energy basis. This cycle could become more attractive as world uranium stocks become depleted, because thorium is approximately four times more abundant than uranium, which combined with the near breeder capability, would result in an almost indefinite supply of fuel. Another attraction of the thorium cycle is that its spent fuel is between 10 and 100 times less radiotoxic than spent fuel from the natural-uranium fuel cycle. The lower radiotoxicity is a result of the absence of uranium-238, the starting point for the formation of transuranic elements, which have very long half lives.

Recovered uranium

This cycle would take uranium recovered from reprocessing operations and use it for CANDU fuel. Reprocessing of spent LWR fuel is conducted in Europe (but not in North America) yielding plutonium and uranium. As discussed in Chapter 7, the plutonium is fabricated into fuel. Currently, nothing is done with the uranium as its content of about 0.9% uranium-235, is too low to be used in LWRs. It is, however, suitable for CANDU reactors, and could be refabricated into useful CANDU fuel, effectively recycling "waste" from light-water reactors. Recycling LWR fuel not only gets much more energy out of the originally mined uranium, but also significantly reduces the amount of waste per unit energy.

Mixed oxide (MOX) fuel cycle

This cycle would use mixed oxide fuel formed by mixing plutonium-

239, as plutonium dioxide, with natural uranium dioxide to form fuel (in the form of ceramic pellets). This is, in fact, the way that plutonium from the European reprocessing plants is currently recycled for LWRs. A smaller concentration of plutonium is needed for CANDU fuel than for LWR fuel. In a similar manner, MOX fuel can be made from plutonium obtained by reprocessing US and Russian military plutonium from decommissioned nuclear bombs. The CANDU is particularly well suited for burning this kind of fuel, and could make a valuable contribution to world disarmament.

The above alternate fuel cycles could be implemented in CANDU reactors with little or no equipment change. Thus, a lasting commitment does not have to be made to any single cycle. The CANDU is the only reactor that provides this flexibility. The ability to use these other fuel cycles also yields a unique synergy between the CANDU and light-water reactors.

ADVANCED CANDU REACTOR DESIGNS

There are several possible paths that can be pursued to further improve the CANDU reactor. The first path involves extensions of the successful CANDU 6 design.

CANDU 3

International marketing studies in the early 1990s showed there was a demand for smaller reactors that would cost less and could be more readily accommodated by smaller utilities. In response, AECL undertook the development of a new CANDU reactor called the CANDU 3, an advanced version of the established CANDU 6, with a nominal power of about 450 MW. Although key components were to be the same as for the existing CANDUs, major advances were introduced in making the new reactor simpler and modular. For example, the use of new technology would reduce wiring by 80%. The CANDU 3 would have two steam generators and two main coolant pumps compared to the four each in the CANDU 6. This would decrease construction time and simplify maintenance. It was also designed to have a longer lifetime of 60 years. With the design about 70% complete, and no reactor orders in view, development effort was halted and transferred to the larger CANDU 9.

CANDU 9

On the large end of the power output scale, the CANDU 9 is a single-station unit of 925-MW capacity based upon the Darlington design, but incorporating some of the development work of the CANDU 3. A key feature is fuel versatility as it can be operated on most of the fuel cycles described in the previous section. The CANDU 9 output could be increased up to 1300 MW by using slightly enriched fuel or by adding more fuel channels. AECL made a proposal to construct two CANDU 9 reactors in South Korea as part of an expansion of that country's electrical power grid. In 2001, it appeared the bid was not successful, and to date no CANDU 9 has been constructed.

NG CANDU

The NG CANDU is a "next generation" advanced reactor now in the conceptual and research phase. It will be heavy-water moderated and light-water cooled. One objective is to significantly reduce the volume (and therefore the cost) of heavy water. A variety of possibilities are under investigation to reduce costs, enhance the engineering process, and include more passive safety systems. For example, a new fuel bundle called CANFLEX® has been developed with 43 fuel elements, instead of the 37 currently used, and has been tested at the Pt. Lepreau reactor. In addition, these elements have two different diameters as well as special appendages to improve the transfer of heat from the bundle into the reactor coolant. Using, for example, the slightly-enriched uranium fuel discussed in the previous section,

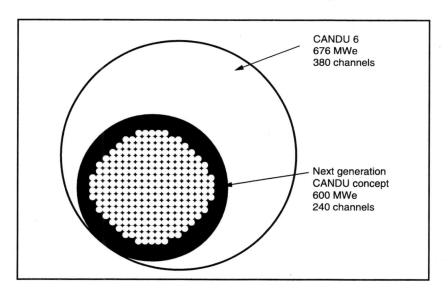

Figure 14-1: An illustration of the reduction in CANDU core size that could be achieved by using slightly-enriched uranium fuel in CANFLEX bundles. The size of the latter configuration is represented by the smaller circle; the larger circle represents today's CANDU 6 reactor (after Torgerson, 2000).

this would mean that more electrical power could be generated in each fuel channel, leading to a smaller reactor core. In turn, many of the other systems could also be scaled down; this is shown conceptually in Figure 14-1. A smaller and simpler reactor means maintenance and reliability would improve, while cost would be reduced. The overall objective is to cut the cost of NG CANDU to about half that of current CANDUs.

Furthermore, new methods of producing heavy water at less cost and avoiding the environmental problems associated with the hydrogen sulphide used in former heavy-water plants are now under development. Figure 14-2 shows a prototype heavy water plant using the new Combined Industrial Reforming and Catalytic Exchange process.

Due to its good neutron efficiency, potential for using advanced fuel cycles, and strong high-technology research support, the CANDU reactor's safety, efficiency, and reliability can be improved, while reducing both its capital and operating costs. In addition, passive systems of the type discussed in the next section for light-water reactors, which can significantly improve reactor safety, are also being researched for CANDU reactors.

THE NEXT GENERATION OF LIGHT-WATER REACTORS

Twenty years have passed since the last new order for a nuclear plant was made by a US utility. The primary reasons have been a flat electricity demand, cost, and the public perception that nuclear reactors are not safe. During this long period, two classes of new light-water reactors have been developed, evolutionary and passive plants, that could alleviate both of the latter two concerns.

Evolutionary Reactors

Evolutionary nuclear plants are advanced versions of existing reactors. The two main evolutionary plants in the United States are an advanced pressurized-water reactor called the System 80+ being developed by BNFL Combustion Engineering and an advanced boiling-water reactor (ABWR) being developed by General Electric. Two ABWRs started operation in Japan in 1996 and 1997, and a design based on the System 80+ has been selected for the next generation of reactors in South Korea. The French and Germans have formed a consortium to develop the evolutionary European Pressurized Water Reactor.

AECL

Figure 14-2: AECL's prototype CIRCE heavy-water plant at Air Liquide in Hamilton Ontario. This new process would 'piggyback' on existing facilities already handling large quantities of hydrogen for other purposes.

To gain economy of scale and, thus, reduce unit power cost, these are all large reactors, about 1,300 to 1,400 MW in capacity. Evolutionary designs incorporate substantial modifications from current plants including: simplified and more rugged systems, more diverse and redundant safety systems, more operator-friendly control rooms, and state-of-the-art computer and data processing technologies. The simplified designs have allowed factory instead of field assembly of many components, reducing overall construction time, which has a significant impact on cost.

The probability of a significant accident has been reduced by more than ten-fold over the current generation of reactors. The simplified design has also yielded a major reduction in construction time; the first two ABWRs in Japan were built in four years each (compared to ten or more years for the previous generation of reactors in North America). Although smaller in power capacity, the CANDU-3 and CANDU-9 discussed in the preceding section are also evolutionary power reactors.

Passive Reactors
Passive plants are dramatically different from the current generation of reactors. They have been designed so that their safety systems

205

respond to accidents automatically, without human or motor intervention. They do this using the natural forces of physics, gravity and convection, and no electricity is required except for what has been stored in emergency backup batteries. The reactor will automatically shut down in the event of an accident, and the decay heat from the fuel will be automatically dissipated without damaging the fuel or reactor core. Fuel melting and, thus, major accidents become virtually impossible. These designs are quite remarkable and should go a long way toward calming the public's fear of catastrophic reactor accidents.

Three types of passive reactors are being developed: light-water cooled, gas-cooled, and liquid-sodium cooled. All rely heavily on the concept of natural convection, that is, as fluids become warmer they expand, become less dense, and rise. Similarly, as fluids become colder they contract, become more dense, and sink. Convection is responsible for chimneys sending smoke and hot air up and out of houses. This convective method of cooling, although untried in a commercial reactor, is used in many research reactors, such as the reactor at McMaster University (see Chapter 16).

Of the three passive light water reactors under development, we will look at the AP600 by Westinghouse as an example (see Figure 14-3). The boiling water version of the passive reactor is called the Simplified Boiling Water Reactor and is being developed by General Electric.

It is estimated that the AP600 can be built in the remarkably short period of three years — this alone make would make it revolutionary. A critical part of reactor design is to ensure that the core can be quickly covered with water to provide cooling should a pipe break and cause the primary coolant to be lost. The AP600 does this without the usual use of pumps, electricity, or operator action by having the water storage tank located inside the reactor containment (instead of outside as at present) and above the reactor core. In an emergency, this water is released and will flow downwards by gravity into the core without requiring pumps or electricity.

Another key aspect is the ability to remove heat in the case of an accident. The AP600 does this in an ingenious manner. In an accident, heat from the reactor fuel elements would cause some of the water covering the core to evaporate. The steam that forms would condense on the walls of the steel containment, if these walls are cooled. Thus, ensuring that the walls are kept cool is essential. As seen in Figure 14-3 the reactor containment consists of a steel shell located

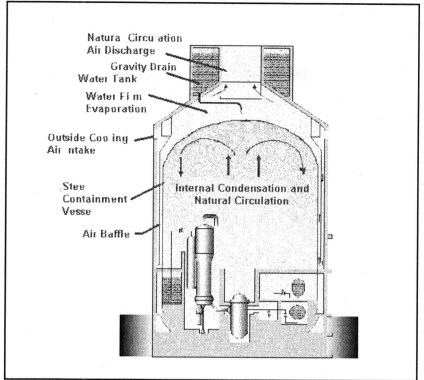

Figure 14-3: The Westinghouse AP600 passive reactor concept is a novel next-generation reactor with significantly increased safety.

inside a concrete building. The building has air inlets as well as a baffle between the building and steel containment vessel which is open at the bottom. Air flows, by natural convection, downward between the concrete building and baffle, makes a u-turn at the bottom and flows upward between the baffle and steel containment to exit at the top of the building. Air is heated as it passes the hot containment shell and rises like smoke in a chimney. At the same time, cold air is drawn from the outdoors to form a natural convection loop without the need for pumps or fans.

To augment the removal of heat, a large tank of water is situated on top of the building whose valves automatically sense a rise in temperature and pressure and release water onto the steel containment shell for the first few days after an accident. The water is vaporized, drawing off additional heat, and the steam flows out of the top of the building. Note that neither the air nor the water that is used for cooling the steel containment shell comes in contact with radioactive materials.

The preceding discussion illustrates only a few of the new and innovative features in the next generation of nuclear passive reactors. Although not described here, these approaches can also be taken with CANDU reactors.

FAST BREEDER REACTORS

These **fast reactors** have no moderator, use enriched uranium or plutonium fuel, and use liquid sodium metal as coolant. The descriptor "fast" refers to the high-energy, or fast, neutrons that arise because there is no moderator to slow them down. These reactors create more fuel as they operate, converting uranium-238 to plutonium-239, hence the name "**breeder**". The central part of the core is enriched to 10 to 20 percent fissile material (either uranium-235 or plutonium-239). Surrounding this is a "breeding blanket" of natural uranium. During operation the uranium-238 in the blanket is irradiated, absorbs neutrons, and transforms to plutonium-239.

This reactor has great promise because not only is uranium-238 abundant and generally of little value (huge stockpiles of depleted uranium, i.e. with the uranium-235 largely removed, have been built up at uranium enrichment plants around the world), but it is also possible to make more fuel than is used. Thus, a virtually inexhaustible supply of nuclear fuel would be available. And this is with technologies that are already in hand, not like fusion or other futuristic technologies that are presently undeveloped.

This type of reactor must be cooled by liquid sodium, a material with low absorption and low moderation of fast neutrons. A drawback to the fast breeder is that sodium leaks and fires have caused problems in several prototype reactors. The most recent was in 1995 at Japan's Monju fast breeder reactor, which suffered a serious sodium leak and has been shut down since then.

Another serious drawback is that plutonium-239 must be extracted from the spent fuel using reprocessing methods that are expensive. With recent discoveries of additional world uranium reserves and the depressed price of uranium, the breeder reactor and the associated reprocessing are not economically competitive at this time, and the priority driving these reactors has diminished. There is also considerable political opposition to extracting plutonium-239, as this is a key material in nuclear bombs.

The US, France, and Japan have experimented with breeder reactors. Canada has not, although the CANDU can be converted into a near-breeder reactor, converting thorium-232 fuel to uranium-233. Generally there is now little enthusiasm for this technology anywhere in the world. Interest could pick up again in the future if an uranium shortage were to arise or if the breeder reactor becomes cost competitive. Many now believe that nuclear fusion, the subject of the next chapter, will be available before the world would need fast breeders reactors.

Chapter Fifteen

Fusion:
The Energy of the Future

Fusion, the energy that powers the sun and the stars, is the ultimate energy of the universe. Most of our present forms of energy are fusion-based, of which solar energy is the most obvious example. The fossil fuels, coal, oil, and natural gas, all originated from ancient plants that grew in sunlight. The sun's energy causes the rains that fill the reservoirs of hydroelectric dams and the winds that turn the blades of wind turbines. All living things on earth owe their existence to fusion.

Humans have long had the dream of controlling fusion, the ultimate power source. With the oceans holding an endless reservoir of hydrogen, the fuel for fusion, we are also driven by the promise of an unlimited supply of cheap fuel. Although progress has been made, the road to harnessing fusion power has been rocky and frustrating. Humans so far have been unable to control a fusion reaction (although uncontrolled fusion has been used in the hydrogen bomb).

FUSION BASICS

Fusion, like fission, is a nuclear process involving interactions between the nuclei of different atoms. Instead of splitting a heavy nucleus, two light nuclei are joined together to form a single nucleus. For some of the light elements, the mass of two individual nuclei is greater than the mass of the combined nucleus. In a fusion reaction, the surplus in mass is converted to energy, according to Einstein's equation. The fraction of matter turned into energy is greater than in fission, making it a process with far more potential for energy generation.

All the light elements up to the mass of iron (atomic number 26) are theoretically capable of undergoing fusion and producing energy. In certain stars, all of these fusion reactions do take place. Under conditions we might produce on earth, however, only the three lightest elements, hydrogen, helium, and lithium have a chance of fusing.

The nuclei of light elements do not join together easily. To convince them to do so, enormous temperatures are required. The fusion reaction that occurs at the lowest temperature, and thus is of most interest, is the deuterium-tritium reaction:

$$^{2}H + \,^{3}H \;\rightarrow\; ^{4}He + n + Energy$$

The same reaction is also written:

$$d + t \rightarrow \alpha + n + Energy$$

In this reaction, "d" and "t" represent the two isotopes of hydrogen, deuterium and tritium, respectively, α represents an alpha particle, or helium nucleus, and "n" stands for a neutron. Written this way it is clear why it is called the "dt reaction."

Deuterium is readily obtained from any source of water since about 150 out of a million hydrogen atoms in nature are deuterium atoms. As a large fusion reactor (1,000 MW) would only use a few hundred kilograms of deuterium per year, a single heavy water plant would supply many dt fusion reactors.

It is proposed that tritium be made by placing a "blanket" of natural lithium in the fusion reactor where it will absorb neutrons to make the required tritium, which can be separated from the lithium by chemical processing. Lithium is a common element in the earth's crust, much more common than uranium. As for deuterium, only a relatively small amount of lithium would be consumed by a fusion power reactor and thus, there are essentially no resource limits on fusion power for far into the future. This widespread availability of fuel is a major attraction of fusion.

212

EINSTEIN'S BRAIN

Albert Einstein, author of the famous equation that describes the relationship between mass and energy, has become the victim of one of the most bizarre episodes ever to befall a great historical figure. The pathologist who performed the autopsy on Einstein after his death in 1956 removed and kept the brain without the authorization of the family. This act resulted in the end of his medical career. He also distributed portions of Einstein's brain to various researchers around the world, and in 1998 transported some larger pieces across the United States by car, as documented in a recent book by Michael Paterniti, *Driving Mr. Albert: A Trip Across America with Einstein's Brain*.

The d-t reaction requires a temperature of about 80 million °C. As this temperature would vaporize all materials, how can a fusion reaction be contained so energy can be extracted to drive turbines and generate electricity?

The answer is to create a plasma, which can be contained by magnetic fields rather than physical walls. By raising the temperature of a gas high enough, the atoms will collide with sufficient energy to remove their electrons. The resulting high-temperature medium, a **plasma**, consists of positively charged ions and negatively charged electrons, whose overall electrical charge is still neutral. Plasma has quite different properties from the gas from which it was formed. For example, a plasma is a very good conductor of electricity because of its free electrons and charged nuclei, whereas the gas it was made from may be a very poor conductor. It is also a good heat conductor, while gases are generally poor in this respect.

Plasma has often been called the fourth state of matter. It is not, in fact, a rare state. It exists on earth in such diverse places as welding arcs, lightning discharges and inside fluorescent-light fixtures. Stars are made up largely of matter in the plasma state. Charged particles from interplanetary space create plasmas by travelling into the earth's atmosphere along its magnetic field lines. The light radiated by these plasmas is called the Northern Lights (see Chapter 3).

Confinement of plasmas at the necessary enormous temperatures is done in two ways. One approach is to confine the plasma in a web (configuration) of magnetic fields which is known as Magnetic Confinement Fusion (MCF). The other main approach to fusion is called Inertial Confinement Fusion (ICF). In this method a great deal of energy is used to create a plasma in a very short time and the fusion process is completed so rapidly that the plasma particles are essentially confined by their own inertia.

INERTIAL CONFINEMENT FUSION

Although most of the international fusion effort is concentrated on magnetic confinement, it is helpful to an overall understanding of fusion to briefly describe inertial confinement.

In inertial confinement fusion, a small pellet consisting of deuterium and tritium is rapidly heated and compressed to a high density by a focussed implosion of the pellet. The objective is to raise the temperature and density high enough so that fusion will be achieved for the short confinement time determined by the inertia of the pellet. Various kinds of high-energy beams, largely lasers but also ion beams, are being investigated to "drive" or produce the implosion of the pellet.

In one method, many intense laser beams are directed simultaneously onto a small spherical fuel pellet from all sides. The pellet may simply be a small shell of glass or plastic filled with a deuterium-tritium mixture perhaps a few millimetres in diameter or an elaborate multi-layered structure designed to promote the implosion and minimize losses. Figure 15-1 shows a demonstration of fusion in such a pellet.

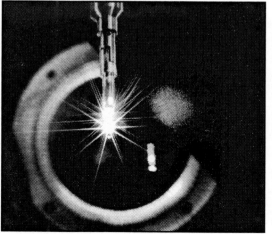

Lawrence Livermore National Laboratories

Figure 15-1: Intense laser beams ignite a small plastic sphere containing deuterium and tritium producing a fusion reaction.

214

The initial intense irradiation of the fuel pellet vaporizes material from the pellet's surface and turns it instantaneously into a corona of plasma blasting away from the still solid pellet. The resulting shock wave compresses the fuel in the pellet to a high density (a thousand or more times liquid density). Laser energy must continue to be absorbed by, or coupled to, the outer corona of plasma in order to continue the compression of the pellet.

One of the basic problems is to achieve efficient coupling of the energy of the lasers to the pellet. It is necessary that the energy be largely used to produce the implosion rather than being dissipated by various non-productive mechanisms. Another important challenge is to construct laser or particle beams of sufficient power, uniformity, energy efficiency, and repetition rate to allow a practical fusion system.

Inertial confinement fusion is perhaps a decade or two behind magnetic confinement fusion in the progress achieved to date. However, the potential for military applications continues to ensure a high level of interest in high-energy lasers. They are used to simulate various aspects of nuclear weapons now that atmospheric weapons testing has been banned. The first major inertial confinement installation, called Shiva, was constructed at Lawrence Livermore National Laboratory in California and uses twenty separate laser beams. An even larger installation called Nova, which is twenty times more powerful, has been in operation there for a decade. Other major ICF facilities are Omega at the University of Rochester and Gekko XII in Japan. Much larger laser ICF installations, the National Ignition Facility in the United States and the Laser Mega Joule project in France, costing in excess of $3 billion US each, are now under construction.

MAGNETIC CONFINEMENT FUSION

Magnetic confinement fusion (MCF) has been pursued continuously since the 1950s and is the method on which most of the world effort is now concentrated. MCF depends on the fact that the electrically charged plasma particles (positively charged hydrogen ions and negatively charged electrons) gyrate around magnetic field lines in a corkscrew fashion. Thus, the plasma can be confined and kept away from the container walls by applying suitably shaped magnetic fields. During this so-called confinement time, the temperature and density of the plasma must be increased to the point where fusion can take place.

It has been rightly said that, "Confining a plasma using magnetic fields is like confining Jell-O with elastic bands." Plasmas are very difficult to confine and decades of research have been required to understand the origins of various instabilities and how to overcome them. There are many possible configurations of MCF devices depending on what arrangements of magnetic fields are used. Some of these systems are illustrated in Figure 15-2. Early researchers contained plasma using cylinders. These were wrapped with electrical conductors through which electricity was passed to create a magnetic field in the cylinder. These open or linear systems initially had great appeal due to their relative simplicity. However, techniques were never found for preventing a significant amount of plasma from leaking out the ends of the cylinder.

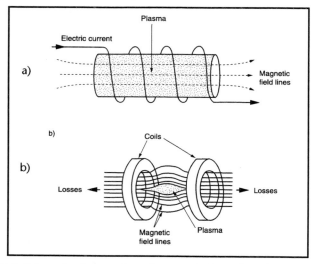

Figure 15-2: Schematic of magnetic confinement. In figure a) an electric current is passed through a wire (or solenoid) wound around a cylinder producing a magnetic field indicated by the dotted lines. In b) using this principle coils of current carrying wire can contain a plasma; however, plasma is lost through the ends. A toroidal (or doughnut) shaped device must be used as shown in Fig 15-3.

This problem was solved by making closed, toroidal (doughnut-shaped) systems. Magnetic coils surround the torus chamber containing the plasma and they create a uniform magnetic field along the axis of the torus. This toroidal field alone, however, is not sufficient to contain the plasma, as it will drift toward the walls of the torus. The addition of another magnetic field, called the poloidal field, in the direction of the small circumference of the torus is required. The resultant spiral field, obtained by adding the two fields, effectively contains the plasma.

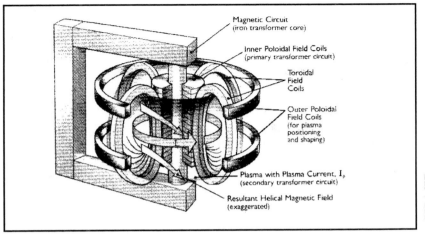

Magnetic Circuit
(iron transformer core)

Inner Poloidal Field Coils
(primary transformer circuit)

Toroidal
Field
Coils

Outer Poloidal
Field Coils
(for plasma
positioning
and shaping)

Plasma with Plasma Current, I_p
(secondary transformer circuit)

Resultant Helical Magnetic Field
(exaggerated)

EFDA-JET

Figure 15-3: Schematic of a tokamak. The magnetic circuit of a transformer induces a current, Ip, in the plasma, which acts as the secondary winding of the transformer. This plasma current produces a poloidal field which together with the toroidal field gives a helical net of magnetic fields that confine the plasma.

Tokamaks

To date, the best device for confining a plasma is the **tokamak**, which was invented in Russia in the 1960s. Tokamak is an abbreviation for a Russian phrase meaning "toroidal magnetic chamber."

A schematic view of a tokamak reactor is shown in Figure 15-3. The toroidal field coils are super-conducting magnets arranged so that a toroidal or doughnut-shaped volume is created. The plasma current circulates like a merry-go-round in this volume and becomes the secondary circuit of a transformer. The primary circuit of this transformer is formed from other poloidal coils placed around the doughnut. The current so created heats the plasma, called inductive heating, but this method alone cannot be used to attain fusion temperatures. The magnetic field from the current itself and the primary circuit coils helps confine and shape the plasma, and to keep it positioned in the toroidal volume. The total magnetic field from the toroidal and poloidal coils and from the plasma itself describes a toroidal helix.

Because of its reliance on transformer action to create the necessary fields, the basic tokamak design is limited to pulsed operation. This means that they create, confine, and heat a plasma for at most a few tens of seconds, at which point the plasma leaks away and the machine must the be prepared for another pulse. Obviously this is an

EFTA JET

Figure 15-4: Exterior and interior views of the JET tokamak, illustrating the size of this device. The volume of the plasma in the Iter machine will be about 15 times larger.

undesirable manner in which to operate an electrical power station. Various non-inductive means for sustaining the plasma current are being explored to see whether steady state, or at least long-pulse operation, is feasible.

A number of tokamaks have been built including the Joint European Torus (JET) in Culham, England, which is operated by the European Community. The scale of the JET machine can be appreciated from Figure 15-4. The Japan Tokamak (JT-60) in Naka, Japan, and the American Tokamak Fusion Test Reactor (TFTR) at Princeton University, New Jersey, which operated from 1982 to 1997, are the other two large tokamaks that have been built. Two other major tokamaks are the DIII-D at General Atomics in San Diego and the Tore-Supra at Cadarache, France. The Canadian research tokamak, the Tokamak de Varennes, near Montreal, operated from 1988 until 1997 when it was shut down and dismantled.

Energy breakeven, a measure of fusion performance, means the fusion energy produced is exactly equal to the energy used to heat the plasma to get the fusion reaction to occur. All three large tokamaks have created plasma conditions with deuterium plasmas that would give energy breakeven if tritium had been used in a dt reaction. JT-60 holds the current record achieving breakeven plus 25 % additional energy. There is a reluctance to use much tritium in JET and TFTR because these machines were not designed to deal with any but a nominal radiological hazard. JT -60 is forbidden by Japanese law to use any tritium at all. Nevertheless, both JET and TFTR ran a few experiments with d-t mixtures, which produced a few tens of megawatts of power.

Other Approaches

In addition to the problems caused by pulsed operation, tokamaks have generally not been very efficient at keeping energy in the plasma. For this reason research on other types of toroidal devices is being pursued throughout the world.

The most developed MCF systems after the tokamaks are stellerators, which use relatively complicated systems of conductors arranged around the torus to create the twisted magnetic fields needed for stable confinement. Since they do not use the transformer effect, stellerators should run continuously instead of being pulsed. A large stellerator called Large Helical Device was completed in Japan in 1998, and a stellerator of another type called Wendelstein VII-X is under construction in the former East Germany. Both of these devices represent investments of about $1 billion US. At this time, stellerators are performing roughly a factor of 10 behind tokamaks in terms of energy breakeven.

Several other types of MCF systems are being pursued at relatively low levels of funding. As yet, however, they are very far from the level of performance of today's tokamaks.

CANADA'S ROLE IN FUSION RESEARCH

Canada was an active and significant contributor to fusion research from the early 1970s until 1997, although our contribution was small compared to that of the major players.

The Canadian Fusion Fuels Technology Project (CFFTP) was located in Mississauga, Ontario, and conducted research based on the unique Canadian expertise in the primary fusion fuels, deuterium and tritium. The studies focused on tritium and fusion fuel systems including tritium blanket studies. Although all of Canada's heavy-water plants are now closed, a large quantity of heavy water has been stockpiled, so that deuterium is readily available. Tritium is produced as a by-product in CANDU heavy water. With 22 large CANDU reactors, Canada is the largest non-military producer of tritium. Furthermore, technology for the safe handling of tritium in terms of worker safety and environmental impact has been extensively developed.

Centre canadien de fusion magnétique

Figure 15-5: The Tokamak de Varennes was Canada's state-of-the-art fusion device located near Montreal. Built at a cost of $100 million, it showed excellent performance until it was scrapped due to government budget cuts in 1998.

Special magnetic confinement and materials studies were conducted at the Tokamak de Varennes from 1988 to 1997 (Figure 15-5).

Both Canadian fusion projects were closed because federal government funding was terminated in April, 1997, due to deficit reduction measures. A small amount of fusion research in still done in Canadian universities, primarily at the Universities of Saskatchewan, Toronto, and Quebec.

THE PROSPECTS FOR FUSION ENERGY

A practical fusion reactor will need more than just the success of the plasma part of the system. A formidable challenge will be to collect the energy and convert it to electricity.

COLD FUSION

On March 23, 1989, two chemists from the University of Utah held a press conference at which they stunned the world by announcing that they had created fusion in a test tube. This experiment, quickly labelled "cold fusion" by the media, became the science story of the decade and set off frantic efforts in other world laboratories to duplicate it. The stakes were very high since this phenomenon was touted by its proponents as the solution to all the world's energy problems.

In the weeks following the announcement, hundreds of laboratories tried the experiment but only a very few claimed to observe any fusion effect. At this time, some chemists tried to become instant experts in fusion physics while some physicists made elementary blunders in chemistry with many scientific reputations damaged by sloppy experimentation and unfounded theorizing. Eventually, it became apparent that cold fusion was an illusion seen only by a small number of enthusiasts. Cold fusion is best remembered as an example of bad science, and a warning to all scientists not to let their enthusiasm blind them to physical realities.

The conversion of plasma energy to electricity will be done via the lithium blanket, which can be in solid or liquid-metal form. The neutrons from the plasma will interact with the lithium giving up kinetic energy and heating the lithium whose coolant (liquid lithium, helium or water) will be pumped through a heat exchanger where it will transfer the heat to steam for driving turbines. There will be major materials problems. New alloys will be needed that can withstand bombardment by high-energy neutrons from the fusion reaction.

Another problem concerns the erosion of the plasma-chamber walls by particles that escape the confinement of the magnetic fields. Wall design is a major challenge, and it will be necessary to replace it periodically during the life of the plant. Of course, the whole reactor will become radioactive because of nuclear reactions initiated by the fusion neutrons and so remote manipulation will be needed to service it.

A fusion reactor will generate radioactive wastes, but these will be smaller in quantity and less radiotoxic than those generated by current fission reactors of similar power. Furthermore, tritium, which is a radiological hazard and requires careful control, will also be involved.

In 1985, the International Thermonuclear Experimental Reactor (ITER) project was initiated jointly by the United States, the European Community, Russia, and Japan. Canada participated through the European Community until 1997. The overall objective of ITER was to demonstrate the scientific and technological feasibility of fusion power and to provide the technical data base for the design and construction of a demonstration fusion power plant. ITER has now been scaled down to a device about half the size and renamed "Iter." Even the scaled-down version is an enormous undertaking and will cost over $6 billion US to construct. In 1998, the United States withdrew from the project, and by 2001 it was still not clear whether Iter would actually be constructed.

Even if Iter is successful, another step would be needed, namely, a demonstration power reactor to ensure that the new materials developed for fusion would actually survive in a working reactor.

The scientific feasibility of fusion energy has been demonstrated in the current generation of large tokamaks. Nevertheless, a great deal of engineering development, lasting for at least 50 years, will be needed for an economic and reliable fusion power system. Therefore, we should not expect to see fusion producing any substantial percentage of the world's energy before the end of this century.

Chapter Sixteen

Research:
The Way Forward

Canadians are more than hewers of wood and drawers of water. Ingenuity and determination have opened up the country through such engineering marvels as the trans-Canada railway that stretches from coast to coast and the St. Lawrence Seaway that allows giant ocean-going ships to sail into the centre of North America. Scientists, engineers, and entrepreneurs have converted our resources into complex, high-technology products. And the more complex and ingenious the product becomes, the more value it gains.

It is the combination of vast resources and creative minds that have brought prosperity to Canada. Our nation is among the world leaders in many high-technology fields such as telecommunications and robotics. Several Canadians have been awarded Nobel prizes in the sciences.

The Canadian nuclear industry is an excellent example of how research and technical innovation can yield new products and economic growth. Nuclear R&D resulted in the unique CANDU reactor that supplies electricity in Canada and in four other countries. Its spinoffs include radioisotopes and cancer therapy machines

AECL

Figure 16-1: Chalk River Laboratories is located on the banks of the Ottawa River about 200 kilometres northwest of Ottawa. The two largest buildings in the centre house the NRX and NRU reactors.

that have transformed the way that medicine is practised. Hundreds of ingenious applications of nuclear gauges and radioisotopes are used in the industrial sector that improve the products we use. And there have been many other advances such as simulator models, computerized control systems, and fundamental progress in metallurgy and chemistry.

THE NATIONAL NUCLEAR LABORATORY

It is only when enough nuclear fuel is assembled, an amount that equals or exceeds the critical mass, that a chain reaction will occur. In an analogous manner, an intellectual critical mass can be created by assembling a large group of top-class scientists and engineers, equipping them with the best facilities available, and concentrating their efforts on a particular area of science and technology. This concept for excellence in research and development is known as a national laboratory and has been used to advance important technology sectors in many countries.

In Canada, there has been a national nuclear laboratory since 1944: the Chalk River Laboratories at Chalk River, Ontario (shown in Figure 16-1). The Chalk River laboratory is a large facility (2,000 people) with a rich history. Chalk River made fundamental contributions to the development of the CANDU reactor starting with the creation of ZEEP in the 1940s and continuing with the NRX, and NRU reactors. Another research facility, Whiteshell Laboratories, also part of AECL, is located at Pinawa, Manitoba. Nuclear research is also carried out at universities and hospitals.

AECL is one of only two government agencies operating major research laboratories in Canada (the National Research Council is the other). The contribution that these laboratories have made to the Canadian economy is difficult to measure, but is certainly significant. The basic and applied research conducted has had spinoffs into many non-nuclear areas.

The discovery of channelling is an excellent example of the interplay between science and applied technology seen in a national laboratory. Nuclear physicists needed to know how far into a solid an ion could penetrate. Determining this quantity, known as the range, theoretically was very difficult since it involves a variety of complex interactions. Therefore, experiments were necessary.

In the early 1960s, chemists in Chalk River used techniques that allowed them to measure very small amounts of radioactive materials dissolved in liquids to measure the ranges of ions in solids. First they implanted radioactive ions in a solid using an ion accelerator. Then they dissolved a very thin layer of the solid. The radioactivity of the dissolved solid layer was measured and the number of radioactive ions in that layer was determined. This process was repeated, peeling off successive layers of the solid. In this way a range profile showing the number of ions that penetrated to each depth was built up. By repeating the experiment for many different ion-solid combinations, the main features of ranges gradually emerged.

The results of most of these painstaking experiments could be explained by existing theories and it looked as if the subject was well understood. In a small percentage of the experiments, however, the ions were penetrating to much greater depths than predicted. As often happens in science, this small anomaly turned out to be extremely important.

At about the same time, scientists at Oak Ridge National Laboratory, one of the national nuclear laboratories in the USA, began computer simulations of this phenomenon using the large high-speed digital computers that had just become available. In today's terms, they set up a virtual solid consisting of an arrangement of atoms and then moved virtual ions through it using Newton's laws of motion to solve the collisions between the ion and the solid atoms. They did this for many ion-solid combinations, effectively performing computer experiments equivalent to the physical experiments going on at Chalk River. The Oak Ridge scientists also observed the anomalously long ranges in a few cases. When the two groups compared results they confirmed that they were observing a new physical phenomenon, not an experimental error.

Figure 16-2 depicts an ion travelling down a "channel", or tunnel, in a solid material. Crystalline solids have regular arrangements of atoms in a three-dimensional pattern called a lattice and if an ion is lined up just right with respect to such a tunnel, it can travel long distances into the solid.

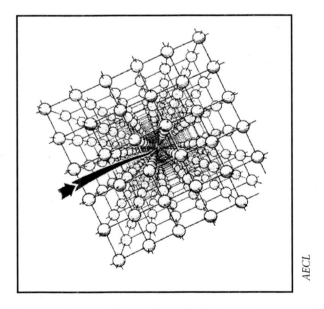

AECL

Figure 16-2: Illustration of channelling showing an ion being steered down a solid lattice.

At first channelling was thought to be just another scientific curiosity. It was soon realized, however, that channelling had very important technological applications. The use of ion implantation to inject trace materials (dopants) into semiconductors was just beginning, and the discovery of channelling initiated a scientific and technological "boom". Scientists from universities and research institutes as well as from semiconductor, telecommunications and computer companies from all over the world came to visit Chalk River and set up collaborative research programs. A new area of science and technology involving hundreds of researchers globally was launched and a series of international conferences was initiated. Applications of ion implantation are not just confined to semiconductors but impact areas such as metal hardening for wear resistance (surface alloying), plasma processing, fusion, and nuclear physics. For example, hundreds of specialized ion implanting accelerators are now used throughout the world to manufacture integrated circuits. Thus, basic nuclear research at Chalk River made a significant contribution to the computer/communications revolution.

This Chalk River story is a good illustration of scientific research: the way that one area of research leads to another, its international aspect, the need for the scientific freedom to pursue the unexpected, and the close relationship between science and technology. It also shows the value of scientists skilled in different disciplines working together in an open intellectual environment and with sound financial support — the ideal situation for a national laboratory.

NUCLEAR SCIENCE

Scientists love firing particles at atoms and studying what happens. A favourite particle is the neutron which, lacking an electrical charge, can pass by the electrons to interact with the nucleus. When a neutron is captured by a nucleus it often upsets the equilibrium, causing a neutron to convert into a proton. To conserve electrical charge, a beta particle is also created and is ejected from the nucleus. The proton stays in the nucleus creating an element with an atomic number one higher than before. For example, a hydrogen atom that gains a proton by this process becomes a helium atom. A helium atom would become a lithium atom and so on.

Uranium is the largest atom found in nature; no natural element contains more protons. Is it possible to create a brand new element heavier than uranium? That is exactly what happened in the 1930s when scientists bombarded uranium with neutrons. They formed an element that no one had ever seen before, with 93 protons, and they named it neptunium, since Neptune is the planet beyond Uranus. And when neptunium decays by beta emission it transforms to a new element with 94 protons called plutonium (Pluto is the planet beyond Neptune). Both neptunium and plutonium are radioactive.

These fascinating developments in nuclear physics were also explored at Chalk River. The first phase of the Tandem Accelerator and Super Conducting Cyclotron (TASCC) was completed in 1987, and full operation began in 1991. It consisted of an accelerator, which would accelerate heavy atoms and then feed them into a cyclotron where the they were further accelerated to almost the speed of light. The cyclotron used magnets made of super-conducting materials, which conduct electricity thousands of times more efficiently than ordinary wires. The building of the superconducting cyclotron, the largest such facility in Canada, was a major technological feat in its time. The accelerated heavy atoms were fired at target nucleii and the interactions were studied. TASCC made many original contributions to nuclear physics before it was closed in 1997 due to federal budget cuts.

Another impact of basic nuclear research facilities at Chalk River was the training of researchers who subsequently moved to the physics and chemistry departments of many Canadian universities. This was particularly pronounced during the large expansion of Canadian universities in the 1960s. Even today, most Canadian physicists have had some past association with Chalk River or one of its offshoots.

Physicists from Chalk River, for example, sparked the building of the Tri University Meson Facility (**TRIUMF**), which started in 1975. TRIUMF (shown in Figure 16-3) is the leading centre for particle physics research in Canada. It is a collaboration between the University of British Columbia, the University of Victoria, Simon Fraser University, and the University of Alberta. This machine uses electromagnets to force positively-charged hydrogen atoms into circular paths at speeds up to 200,000 kilometres per second (67% of the speed of light). These ion beams are then shot onto special targets to study the subatomic fragments that result from the collision. Exotic subnuclear particles such as mesons, pions, muons, and neutrinos are formed that only theoretical physicists can love or understand. TRIUMF also produces radioisotopes that are deficient in neutrons such as cobalt-57, gallium-67, indium-111. These are packaged and marketed by MDS Nordion.

The Sudbury Neutrino Observatory (**SNO**), another basic-science facility involving graduates of Chalk River, began operating in 2000. Built deep in the Creighton nickel mine in Sudbury, its purpose is to measure the neutrinos emitted by the sun and other stars. The number and the type of neutrinos indicate the nature of the nuclear

TRIUMF

Figure 16-3: TRIUMF, shown during construction, is the world's largest cyclotron at 17 metres in diameter.

reactions taking place in stars and give scientists insights on the evolution and composition of the universe. Neutrinos are very difficult to observe since they tend to pass through material objects. The observatory is located deep in a former nickel mine where energetic cosmic-ray particles cannot penetrate and interfere with the experiment. The detector is an acrylic plastic sphere filled with 900 tonnes of heavy water surrounded by ordinary water. Highly sensitive light detectors surround this sphere in a totally dark cavern and detect the photons emitted in the rare events when a neutrino hits one of the nuclei in the heavy water.

In June 2001, the SNO team announced that their observations confirmed the neutrino had a mass, a result first reported in 1998 from Super-K, a similar Japanese experiment. They also observed all of the types of neutrinos emitted by the sun, something not possible at Super-K, from which they concluded the sun's output of neutrinos was in excellent agreement with theoretical predictions. These announcements were made to international acclaim because of their profound signifigance to fundamental physics and our understanding of the universe.

Bertram Brockhouse (1918-): was born in Alberta and educated at the Universities of British Columbia and Toronto. From 1950 to 1962, he made ground-breaking discoveries in the physics of solids using neutrons from Chalk River's NRX and NRU reactors, in the process developing a new instrument for neutron scattering, the triple-axis spectrometer, which for the first time permitted the measurement of the changes in energy of neutrons after they have moved through matter. He was the first to measure the energy-versus-momentum (or the frequency-wavelength) relationship for lattice waves in crystals, liquids, and magnetic materials. He received many honours including Fellowship in the Royal Society of London, the Order of Canada, and the joint award of the Nobel Prize in physics in 1994. Now he is Professor Emeritus at McMaster University, which he joined in 1962.

AECL

Neutrons, according to quantum physics theory, have the properties of both a particle and a wave. For thermal neutrons (i.e. with low energy) the wavelength happens to be similar to the spacing of atoms in a crystal, allowing them to be diffracted by a crystal in the same way that light is diffracted by a prism or lines on a grating. Thus, neutrons can be used to study the structure of atoms.

INNOVATION FROM NUCLEAR RESEARCH

Previous chapters have described the impact nuclear technology has had on energy, medicine, industry, and science. Here we will describe some specific examples of the innovations that have flowed from the Canadian nuclear research and development program.

Research into the effects of radiation on human health has been conducted at Chalk River for over 35 years. It was at Chalk River that the ideas for cancer therapy using cobalt-60, positron emission tomography, and the thermoluminescent badge for personal dose monitoring originated. Research into molecular biology and mechanisms of cancer induction have made significant contributions in the campaign to find a cure for cancer. Recently, a method has been developed that may indicate the likelihood of people to develop cancer. In 2001, these research areas were in the process of being transferred from AECL to Health Canada.

From its outset, Chalk River has conducted research into the use of radionuclides for industry and medicine, as well as producing radionuclides in its reactors. Some years ago Nordion was spun out of AECL as an independent business. It was subsequently brought by MDS, renamed MDS Nordion, and has become an international leader in the medical and industrial application of radiation. In the 1990s, MDS Nordion funded AECL to design and build the world's first dedicated reactors for producing radioisotopes. These two MAPLE reactors are being commissioned at Chalk River, which provides the necessary nuclear infrastructure for them. The radioisotopes will be sold worldwide by MDS Nordion.

Chalk River and Whiteshell were at the forefront of developing accelerators, which are widely used in research and have begun to replace gamma-ray machines for cancer therapy and sterilization of medical supplies. Accelerators are not nuclear devices; they are electrically driven and produce beams of high-energy particles such as electrons, which are used for a large variety of industrial processes including production of heat-shrinkable plastics, printing of beer-can

TRAINED SEALS

In the early days of the CANDU program, pump seals in the primary coolant circuit of the reactors had to be replaced at least once a year because of the high pressure and temperature conditions in which they operated. To avoid these lengthy and costly reactor shutdowns, a research program was initiated to develop seals that would last at least five years. Using computer modeling as well as extensive laboratory testing, Chalk River staff gained a fundamental understanding of the mechanisms and material properties involved, and developed a seal that met the objectives. Power utilities across North America were experiencing similar problems, and in the late 1970s and early 1980s Chalk River won $18 million in contracts to solve pump-seal problems for three US nuclear electric utilities.

But the highlight of this research came after the tragic explosion of the US space shuttle Challenger in 1986 in which seven lives were lost. The cause was traced to hydrogen fuel leaking past an O-ring seal at the base of the booster rocket. The development of new O-rings was given top priority by NASA, and AECL was hired as a specialist subcontractor to advise on the re-design and to perform tests. In recognition of their efforts, two Chalk River scientists were invited to attend the launch of the first space shuttle with the new O-ring. No problems have been encountered since that time.

labels, and sterilization of medical supplies. Current accelerators are restricted by their limited power to surface coatings or the processing of thin sheets of material (typically a few millimetres). IMPELA (Industrial Materials Processing Electron Linear Accelerator) was developed with a much higher power that allows it to penetrate more deeply and to handle larger quantities of materials.

The market for accelerators is growing rapidly. Potential future applications include processing of natural fibre such as wood products to alter the size of molecular chains, for example, altering wood fibre into cellulose-reinforced plastics. Electron beams can also be used in the pulp and paper industry to improve efficiency and reduce the use of chemicals. Some of AECL's accelerator activities were recently transferred to a new private company, Acsion Industries.

Chalk River and Whiteshell also became international centres of research on zirconium and its alloys, materials that are vital to CANDU reactors.

Chalk River has tried to augment its shrinking government funding with non-nuclear commercial research work. The program has had a few successes, including a high profile O-ring design for the American space-shuttle program (see sidebar). Another high-technology application is the detection of metal particles in motor oil using non-intrusive eddy-current tests. This allows the oil in the engines of car and truck fleets to be readily tested so they can be serviced when necessary.

AECL's Whiteshell Laboratories was established in 1962 in eastern Manitoba and in its hey-day had a staff of about 1200. A large research program was carried out, much of it in collaboration with Chalk River. Whiteshell housed the WR-1 reactor, a novel organic-cooled, heavy-water-moderated reactor, which could operate at higher temperatures (350 to 400 $^{\circ}$C) and thus better efficiency, than the water-cooled CANDU reactor (300 $^{\circ}$C). However, the 60 MWt reactor, which used slightly enriched uranium fuel and was commissioned in 1965, was shut down in 1985 and the research discontinued when the CANDU proved to be successful and budgets became tight.

Canada's research into the deep underground disposal of spent nuclear fuel from reactors was coordinated from Whiteshell. A major focus is at a nearby rocky granite outcrop, where the shaft for the Underground Research Laboratory penetrates into the Canadian

Shield. A series of experiments have been undertaken to simulate the deep disposal of spent nuclear fuel (see Chapter 10). An inherently safe reactor, the **SLOWPOKE** (Safe Low Power Critical Experiment), was studied at Whiteshell as a prototype for a 10 MWt heat-producing unit for use in district heating. Although the concept had some technical merit, it was abandoned because of lack of market interest.

In 1998, it was announced that Whiteshell will be privatized by the year 2001. It is currently operating with a pared-down staff of about 250. It will be interesting to see if privatization will be successful in transforming science into commercial products and services.

NUCLEAR RESEARCH IN THE UNIVERSITIES

A number of Canadian universities conduct nuclear research using small research reactors, primarily the SLOWPOKE (see Figure 16-4).

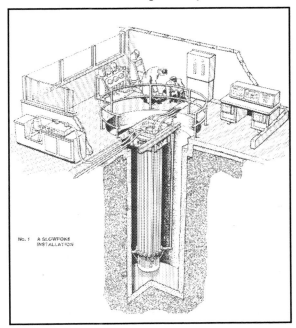

No. 1   A SLOWPOKE
        INSTALLATION

AECL

Figure 16-4: Schematic of a SLOWPOKE II reactor showing its simplicity. This highly versatile research reactor can operate unattended for long periods of time.

Research reactors have a wide range of uses, including:

• neutron scattering in which beams of thermal neutrons are scattered by atoms in a target, yielding information regarding the atomic structure, magnetic state, and atomic binding energies of the target material;

233

- neutron activation analysis, a method for detecting very low concentrations of elements in a sample;

- neutron radiography (the neutron analogue of X-Rays);

- irradiation testing of materials;

- production of radioisotopes for research, medicine, and industry.

These reactors are used by researchers in an incredibly broad range of fields, such as archaeology, materials science, fusion research, mineral prospecting, and environmental science, to name a few.

Research reactors are generally much smaller and simpler than power reactors with power levels usually only a fraction of one percent of a typical CANDU reactor. Their cores can often have very high power density, although contained in a very small volume.

Table 16-1
CANADIAN UNIVERSITIES WITH REACTORS

University	Type	Thermal Energy
McMaster University	Swimming Pool	5000 kW
University of Toronto	SLOWPOKE II	20 kW (closed 1998)
Royal Military College	SLOWPOKE II	20 kW
Saskatchewan Research Council	SLOWPOKE II	20 kW
University of Alberta	SLOWPOKE II	20 kW
Ecole Polytechnique	SLOWPOKE II	20 kW
Dalhousie University	SLOWPOKE II	20 kW

The SLOWPOKE reactor has become an important research tool, beginning in 1971 when the first one was licensed at the University of Toronto. Four SLOWPOKEs are currently in operation at Canadian universities, one at the Saskatchewan Research Council, and another at the University of West Indies in Jamaica. The SLOWPOKE at the University of Toronto was closed in 1998.

The key feature of the SLOWPOKE is that it is inherently safe and can run unattended. It is a small reactor with a fuel assembly about the size of a 4-litre can of paint. The core is located at the bottom of a cylindrical tank of ordinary water. The fuel consists of highly enriched uranium (about 90% uranium-235) that generates only a small amount of heat, about 20 kilowatts. Currently, the highly

THE DALHOUSIE REACTOR

The SLOWPOKE reactor at Dalhousie University in Halifax, Nova Scotia, has thrust its Trace Analysis Research Centre into the forefront of research in analytical chemistry. Installed in 1976, the reactor is used for Neutron Activation Analysis. In this method a sample is irradiated in the reactor, where the nuclei absorb some neutrons, become unstable, i.e. radioactive, and emit gamma rays, which are measured with a gamma-ray detector. This yields a gamma energy spectrum, which can be compared to the spectra of known elements to identify the elements in the sample.

It is imperative that the samples be removed from the reactor and placed in a detector before the elements with very short half lives decay away. The scientists at Dalhousie have developed a system which does the transfer in 0.35 seconds. The reactor is used to study, among many other things, how trace elements are bound to proteins in food and in the body. The SLOWPOKE is also used to produce isotopes for medical research.

enriched fuel in all these research reactors is being replaced by low-enriched uranium, about 20% uranium-235. The small size of the core is possible because it is surrounded by beryllium, which reflects neutrons back into the core. The heated water rises to the top of the tank where it loses its heat by flowing through a heat exchanger. The cool water sinks to the bottom of the tank, completing the natural convection cycle.

H. G. (Harry) Thode (1910-1997), a pioneering nuclear chemist, graduated from the University of Saskatchewan and did graduate work at the University of Chicago. In the early 1940s, he built the first mass spectrometer in Canada, which he used to determine isotope ratios. Over the course of a long career he made many distinguished contributions to nuclear science and to education. He was responsible for the construction of the nuclear reactor at McMaster University, then the first university reactor in the British Commonwealth. He served as President of McMaster University and among many honours was one of the first 10 Canadians to become a Companion of the Order of Canada.

The SLOWPOKE is very simple in design as no pumps are needed and the water is not pressurized as in a CANDU reactor. Because the depth of water provides good shielding, the top can be kept open to allow easy access for monitoring, equipment, and experiments.

Due to the inherent passive safety, no elaborate, fast-acting shut-down systems are needed. As the temperature around the core rises, the water becomes less dense and is less efficient as a moderator, automatically causing the nuclear reaction to slow down. Thus, any extra heat acts as a regulator and slows down the reaction. The SLOWPOKE has been designed so that this self-regulation can cope with all possible power variations, making it truly safe to operate even in large population centres. SLOWPOKE is one of the ancestors of today's MAPLE series of reactors.

The McMaster Nuclear Reactor, the oldest and largest university reactor in the British Commonwealth, was opened in 1959 at McMaster University in Hamilton, Ontario. Since then, it has provided facilities for neutron beam experiments, isotope production, neutron activation research, neutron radiography research, and education in many fields including materials science, nuclear science and engineering, and health and radiation physics. The reactor is a 5 MWt pool-type reactor of the Materials Testing Reactor design (the core is shown on the cover). Although relatively common in the USA, this is the only such reactor in Canada. It provides about 250 times more power than the SLOWPOKE reactor used at other Canadian universities.

The McMaster reactor contributes to aircraft safety by performing radiographs of high-temperature jet engine turbine blades to detect structural flaws. It also produces about 25% of the world's supply of iodine-125, an isotope used in nuclear medicine for the treatment of prostate cancer. The fuel, which consists of uranium-aluminum plates enriched to about 20% uranium-235, is clad in aluminum and sits in a 10-metre deep pool filled with light water.

In 1999, the Canadian Foundation for Innovation, the Ontario Innovation Trust, industry, and the university together contributed $10 million to refurbish the McMaster reactor and support new research programs. This capital injection, together with the revenues from its commercial isotope production, will ensure its continued operation for many years.

THE NEED FOR NEUTRONS

In the late 1990s, governments in Canada recognized that research facilities in universities were run down and needed upgrading. They also recognized that research and the innovative technologies it produces are the key to the growth of Canada's increasingly knowledge-based economy. Therefore, billions of dollars are being invested in improving research infrastructure in Canadian universities. In parallel, up to 2,000 new research professorships are being funded. These programs are producing an upsurge of activity in research of all kinds.

Nuclear R&D in Canada, however, has been in decline since the halcyon days of the sixties and seventies. This decline became precipitous during the mid-1990s when deficit reduction measures greatly reduced federal funding at the same time as nuclear utilities decreased their research budgets. Thus, recruiting new graduates for the nuclear industry has become increasingly difficult at a time when retirements are rapidly depleting the work force.

The closure of the NRU reactor, scheduled for 2005, will leave a major gap in Canada's nuclear research capability. The existing university reactors cannot provide the high neutron fluxes and research infrastructure available at a national laboratory. Therefore, AECL and the National Research Council, in partnership with universities and industry, submitted a proposal in 1999 to replace the aging NRU reactor with a new research reactor called the Canadian Neutron Facility. It would provide advanced materials research capability for universities and industry, helping maintain Canada's competitiveness in this field. It would also provide a testing facility to support and advance CANDU reactor development. In addition, it would provide the focal point for a new generation of nuclear researchers.

Neutrons produced by reactors are a unique, deep-penetrating, and non-destructive probe that can be used to study a variety of materials including pharmaceuticals, magnetic devices such as computer disks, polymers, welded structures, superconducters, biological materials and much more. In addition, only neutrons can detect residual stress deep inside alloys and ceramics, and so contribute to the reliability and safety of new industrial products.

About one hundred Canadian and international scientists and engineers, as well as about 20 graduate students, currently conduct research at AECL's NRU reactor. This number is expected to more than triple if the Canadian Neutron Facility comes on line.

To help put the Canadian Neutron Facility in context, it can be compared to another research facility, the Canadian Light Source, due to be completed in 2002 near Saskatoon, Saskatchewan. This is a type of electron accelerator called a synchrotron that accelerates a fine beam of electrons to very near the speed of light in a race-track-shaped accelerator about the size of a football field. At such high speeds, very bright x-ray beams are produced. These beams will be used to probe a variety of materials including biological molecules and tissues. The Canadian Light Source is a necessity for modern materials research and Canada was the only member of the G-8 countries without one.

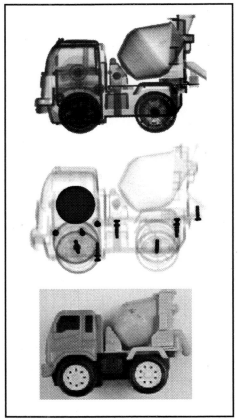

NRay Services Inc.

Figure 16-5: Imaging a toy truck using normal photography (bottom panel), neutrons (top panel), and X-rays (middle panel). Note that the X-rays show metal components very well while the neutrons reveal other structures not seen by X-Rays.

The simple case of the toy truck illustrated in Figure 16-5 shows why it is also essential for every G-8 country to have a modern neutron source. The reason is that neutrons and synchrotron radiation "see" materials differently, thus complementing each other. Researchers need both, as it is most effective to investigate the various aspects of materials using both neutrons and X-rays. Australia, for example, recognizes this and is building a new research reactor and a synchrotron side by side. Many countries, such as Australia, that do not have nuclear power programs operate research reactors because of the enormous application to science and technology.

To continue to advance innovation in many areas, Canada also will need a modern neutron source or we will again find ourselves behind other countries. The Canadian Neutron Facility would also help ensure the survival of a national laboratory whose ingenuity will continue to transform R&D into innovative technological applications that benefit all Canadians.

Bibliography

Atomic Energy of Canada Limited. 1994. Environmental Impact Statement on the Concept for *Disposal of Canada's Nuclear Fuel Waste*. AECL-10711, COG-93-1.

Atomic Energy of Canada Limited. 1997. *Canada Enters the Nuclear Age, A Technical History of AECL*. Montreal and Kingston: McGill-Queens University Press.

Atomic Energy Control Board. 1995. *Canada: Living with Radiation*. Ottawa: Canada Communication Group.

Atomic Energy Control Board. 1998. *Radioactive Emission Data from Canadian Nuclear Generating Stations 1988 to 1997*. INFO-0210/Rev.8, Ottawa.

Ball, N. R. 1987. *Professional Engineering in Canada 1887 to 1987*. National Museum of Science and Technology/National Museums of Canada, Ottawa.

Blix, H. 1997. Nuclear Energy in the 21st Century. *Nuclear News*. pp.34-39, September.

Bolton, R. A. 1999. Fusion Research's Demise in Canada: another Avro Arrow? *Canadian Nuclear Society Bulletin*. vol. 20, no. 3, pp. 43-46.

Bothwell, R. 1984. *Eldorado: Canada's National Uranium Company*. Toronto: University of Toronto Press.

Bothwell, R. 1988. *Nucleus: The History of Atomic Energy of Canada Limited*. Toronto: University of Toronto Press.

Bushby, S. J., Dimmick, J. R. and Duffey, R. B. 1999. Conceptual Designs for very High Temperature CANDU Reactors. *Canadian Nuclear Society Bulletin*. vol. 21, no. 2, pp. 25-31.

Calaprice, A. 1996. *The Quotable Einstein*. Princeton, N. J.: Princeton University Press.

Canadian Environmental Assessment Agency. 1998. *Nuclear Fuel Waste Management and Disposal Concept*. Report of the Environmental Assessment Panel on the Nuclear Fuel Waste Management and Disposal Concept. Ottawa.

Carbon, M. W. 1997. *Nuclear Power: Villain or Victim? Our most misunderstood source of electricity.* Madison, Wis.: Pebble Beach Publishers.

Cohen, B. L. 1990. *The Nuclear Energy Option – An Alternative for the 90s.* New York, N.Y.: Plenum Press.

Cohen, B., L. 1995. Test of the Linear No-Threshold Theory of Radiation Carcinogenesis for Inhaled Radon Decay Products. *Health Physics,* vol. 68, no. 2 pp.157-174.

Cole, H. A. 1988. *Understanding Nuclear Power.* Aldershot, U. K.: Gower Technical Press.

Coursey, B. M. and Nath, R. 2000. Radionuclide Therapy. Physics Today, April, pp. 25-30.

Cousins, T. et al. 2000. Using Thermal Neutron Activation to Detect Nonmetallic Land Mines. *Canadian Nuclear Society Bulletin,* vol. 21, no. 2, pp. 40-44.

Davies, J. A., Amsel, G., and Mayer, J. W. 1992. Reflections and Reminiscences from the Early History of RBS, NRA and Channeling. *Nuclear Instruments and Methods in Physics Research.* Section B, vol. 64, pp. 12-28.

Duport, P. 2000. The Effects of Low Doses of Ionizing Radiation. *Canadian Nuclear Society Bulletin.* vol. 21, no. 3, pp. 20-23.

Eisenbud, M. 1968. Sources of Radioactivity in the Environment, Proceedings of a Conference on the Pediatric Significance of Peacetime Fallout, *Pediatrics,* supp. 41, pp. 174-195.

Eisenbud, M. 1973. *Environmental Radioactivity.* 2nd edn. New York, N.Y.: McGraw Hill.

Eldorado Resources Limited. 1985. *Uranium and Electricity.* 5th edn. Ottawa.

Ernst & Young. 1993. *Study of the Economic Benefits of the Canadian Nuclear Industry.* Conducted for Atomic Energy of Canada Limited.

Foster, John. 2000. In memoriam Harold Smith, *Canadian Nuclear Society Bulletin.* vol. 20, no. 4, pp. 43-45.

Fuller, J. G. 1975. *The Day We Almost Lost Detroit,* New York, N.Y.: Reader's Digest Press.

Gray, C. 1984. Nuclear Energy & Medicine: The Canadian Connection, *Canadian Medical Association Journal,* vol. 130.

Harms, A. A. , Schoepf, K. F., Miley, G. H., and Kingdon, D. R. 2000. *Principles of Fusion Energy*. Singapore: World Scientific.

Hart, R. S. 1997. *CANDU Technical Summary*, Atomic Energy of Canada Limited.

Hedges, K. R. and Yu, S. K. W. 1998. Next Generation CANDU Plants. *Canadian Nuclear Society Bulletin*, vol. 19, no. 3, pp. 16-20.

Hincks, E. P. (ed.) 1979. Nuclear Issues in the Canadian Energy Context, *Proceedings of a National Conference, Vancouver, March 7-9, 1979*, Royal Society of Canada and the Science Council of Canada, Ottawa.

Hore-Lacy, I. 1997. *Nuclear Electricity*, 4th edn., Melbourne, Australia: Uranium Information Centre Ltd.

House of Commons, Canada. 1988. *Nuclear Energy: Unmasking the Mystery*. 10th report, Standing Committee on Energy, Mines and Resources, Ottawa.

Hoyle, F., and Hoyle, G. 1980. *Commonsense in Nuclear Energy*. San Francisco, Calif.: W. H. Freeman and Company.

Hurst, D. G. 1989. The Road to CANDU, *Canadian Nuclear Society Bulletin*, vol.10, no.4, pp.17-21.

Ing, H. et al. 1996. Bubble Detectors and the Assessment of Biological Risk from Space Radiations, *Radiation Protection Dosimetry*, vol. 65, pp. 421-424.

International Atomic Energy Agency. 1997. *Ten Years After Chernobyl: What Do We Really Know?*, summary of the Proceedings of the IAEA/WHO/EC International Conference in Vienna, April 1996. Vienna.

International Atomic Energy Agency. 1999. *The International Nuclear Event Scale*, Vienna.

International Atomic Energy Agency. 2000. Nuclear Power Status Around the World. *IAEA Bulletin*. vol. 42, no. 3, p. 70.

Jackson, D. P. 1981. Three Mile Island: a Personal Commentary. *Energy Newsletter (The McMaster Institute for Energy Studies)*. vol 1. pp. 4-13.

Jackson, D. P. and de la Mothe, J. 2001. Nuclear Regulation In Transition: The Atomic Energy Control Board At The Turn Of The Century. *Proceedings of the Conference on the Future of Nuclear Energy in Canada*. Toronto: University of Toronto Press.

Jones, N. 2001. The Monster in the Lake. *New Scientist*, March 24, 2001 pp. 36-40.

King, A., and Schneider, B. 1991. *The First Global Revolution, A Report by the Council of the Club of Rome,* New York, N.Y.: Pantheon Books.

Lambert, B. 1990. *How Safe is Safe? Radiation Controversies Explained,* London, U. K.: Unwin Paperbacks.

Lau, J.H.K. et al. 1999. The Canadian CANDU Fuel Development Program and Recent Fuel Operating Experience. *Canadian Nuclear Society Bulletin,* vol. 20, no. 3, pp. 8-15.

Lidstone, R. F. 1996. The development of MAPLE Technology. *Canadian Nuclear Society Bulletin,* vol. 17, no. 4, pp. 32-38.

Litt, P. 2000. *Isotopes and Innovation, MDS Nordion's First Fifty Years, 1946-1996.* Montreal and Kingston: McGill-Queens University Press.

Lochbaum, D. 1998. *The Good, the Bad and the Ugly, A Report on safety in America's Nuclear Power Industry,* Cambridge, Mass: Union of Concerned Scientists.

Marmorek, J. 1978. *Everything You Wanted to Know About Nuclear Power (But Were Afraid to Find Out!),* Toronto: Energy Probe and the Pollution Probe Foundation.

McHughen, A. 2000. *Pandora's Picnic Basket, The potential and hazards of genetically modified foods.* Oxford U. K.: Oxford University Press.

McNeill, J. R. 2000. *Something New Under the Sun: An Environmental History of the Twentieth Century World.* New York, N.Y.: W. W. Norton & Company Inc.

Miller, Jr., G. T. 1997. *Environmental Science – Working with the Earth,* Belmont, Calif.: Wadsworth Publishing Company.

Moore, B. and Guindon, S. 1999. Competitiveness of Nuclear Energy. *CRUISE Conference on the Future of Nuclear Energy.* Toronto: University of Toronto Press.

Morrison, R. W. 1998. *Nuclear Energy Policy in Canada: 1942-1997.* Carleton Research Unit on Innovation Science and Environment. Ottawa: Carleton University.

Murray, R. L., 2001. *Nuclear Energy.* Boston, Mass.: Butterworth Heinemann.

National Research Council. 1990. *Health Effects of Exposure to Low Levels of Ionizing Radiation: BEIR, V,* Committee on the Biological Effects of Ionizing Radiation, Washington, DC.: National Academy Press.

National Research Council. 1999. *Condensed-Matter and Materials Physics: Basic Research for Tomorrow's Future.* Washington, D. C.: National Academy Press.

Natural Resources Canada. 1997. *Canada's Energy Outlook: 1996-2020.* Ottawa.

Natural Resources Canada. 2000. *Energy in Canada 2000.* Ottawa.

Nuclear Energy Agency. 1995. *The Environmental and Ethical Basis of Geological Disposal of Long-lived Radioactive Wastes.* Paris.

Nuclear Energy Agency. 1998. *Nuclear Power and Climate Change.* Paris.

Nuclear Energy Agency. 2000. *Nuclear Education and Training: Cause for Concern?* Paris.

Ontario Hydro. 1988. *A Journalist's Guide to Nuclear Power.* Toronto.

Organizations United for Responsible Low-Level Radioactive Waste Solutions. 1994. *The Untold Story: Economic and Employment Benefits of the Use of Radioactive Materials.*

Paterniti, M. 2000. *Driving Mr. Albert, A Trip Across America with Einstein's Brain.* New York, N.Y.: Dial Press.

Pearce, F. 2000. Kicking the [Petrol] Habit. *New Scientist,* November 25, 2000. pp. 34-40.

Perrow, C. 1984. *Normal Accidents, Living With High Risk Technologies.* New York, N.Y.: Basic Books.

Poch, D. I. 1985. *Radiation Alert.* Toronto: Energy Probe/ Doubleday Canada Ltd.

Pool, R. 1997. *Beyond Engineering, How Society Shapes Technology.* Oxford, U. K.: Oxford University Press.

Reynolds, A. B. 1996. *Bluebells and Nuclear Energy.* Madison, Wis.: Cogito Books.

Rhodes, R. and Beller, D. 2000. The Need for Nuclear Power. *Foreign Affairs.* vol. 79, no. 1, pp. 30-44.

Royal Society and Royal Academy of Engineering. 1999. *Nuclear Energy, the future climate.* London, U. K.: Royal Society document 11/99.

Seligman, H. 1990. *Isotopes in Everyday Life,* International Atomic Energy Agency, Vienna.

Shapiro, J. 1981. *Radiation Protection: A guide for Scientists and Physicians,* 2nd edn. Cambridge, Mass.: Harvard University Press.

Shcherbak, Y. M. 1996. Ten Years of the Chornobyl [sic] Era, *Scientific American,* April 1996, p. 44- 49.

Sims, G. 1981. *A History of the Atomic Energy Control Board.* Ottawa: Canadian Government Publishing Centre.

Sims, G. 1990. *The Anti-Nuclear Game.* Ottawa: University of Ottawa Press.

Snell, V. G. and Howieson, J. Q. 1991. *Chernobyl – a Canadian Perspective,* Atomic Energy of Canada Limited report 910580PA, revised.

Snow, C. P. 1964. *The Two Cultures: And a Second Look.* Cambridge, U. K.: Cambridge University Press.

Snow, C. P. 1981. *The Physicists.* Boston, Mass.: Little, Brown and Company.

Stacey, W. M. 2001. *Nuclear Reactor Physics.* New York, N.Y. Wiley and Sons.

Starfelt, N. 1999. *Accounting for Environmental Costs in the Production of Electricity.* American Nuclear Society Conference, Long Beach, 1999 Nov. 14-18.

Steane, R. 1997. Uranium Update, *Canadian Nuclear Society Bulletin.* vol. 18, no. 1, pp. 28-31.

Tammemagi, H. and Smith, N. L. 1974. A radiogeologic study of the granites of south-west England. Journal of the Geological Society, vol. 131, pp.415-427.

Tammemagi, H. 1999. *The Waste Crisis, Landfills, Incinerators, and the Search for a Sustainable Future.* New York, N.Y.: Oxford University Press.

Torgerson, D. F. 2000. Reducing the Cost of CANDU. *Canadian Nuclear Society Bulletin.* vol. 20, no 4. pp. 22-25.

United Nations Scientific Committee on the Effects of Atomic Radiation. 1988. *Sources, Effects, and Risks of Ionizing Radiation.* United Nations Scientific Committee on the Effects of Atomic Radiation. New York, N. Y.: United Nations.

US National Committee on Radiological Protection and Measurements. 1980. *Influence of dose and its distribution in time on Dose-relationships for low LET radiation,* NCRP Report No. 64.

Whitlock, J. 1994. McMaster Nuclear Reactor Turns 35. *Canadian Nuclear Society Bulletin,* vol. 15, no. 1, pp. 2-5.

Zeyher, A., 1997. Targeted Isotope Homes in on Hodgkin's Disease, *Nuclear News,* pp. 58-62, October 1997.

246

Glossary

Absorption: A type of nuclear reaction in which a nucleus captures a neutron.

Abundance: In the context of an isotope, the relative occurrence of that particular isotope compared to others of the same element.

Actinides: A group of 15 elements with atomic number from 89 (actinium) to 103 (lawrencium). All are radioactive.

AECB: Atomic Energy Control Board, the former name of the CNSC (see below).

AECL: Atomic Energy of Canada Limited, the federal crown corporation responsible for nuclear research and development, and CANDU reactor sales and service.

ALARA: As Low As Reasonably Achievable, a philosophy of nuclear safety in which risks should be maintained not just at regulatory limits but even lower, as reasonably achievable, economic and social considerations being taken into consideration.

Alpha particle (α): A positively charged particle emitted in the radioactive decay of certain radioactive atoms. An alpha particle is identical to the nucleus of a helium atom, consisting of two neutrons and two protons.

Atom: The basic building block of matter. It is the smallest unit of a material that retains all of the properties of that material. It consists of a nucleus surrounded by electrons.

Atomic number: The number of protons in the nucleus (and the number of orbital electrons) of an atom. This determines which chemical element an atom is. For example, an atom with six protons is carbon.

Becquerel (Bq): The unit for expressing the rate of radioactive decay. One becquerel is equivalent to one disintegration per second.

Beta particle (β): A negatively charged particle that is emitted by certain radioactive (unstable) atoms; a beta particle is an electron that originates in the nucleus.

Breeder reactor: A reactor that produces more fissile material (fuel) than it consumes.

Burnup: The amount of energy produced by a given amount of nuclear fuel.

BWR: Boiling Water Reactor. A type of light water reactor that generates steam directly in the reactor vessel.

Calandria: The vessel that forms the core of a CANDU reactor. It contains the heavy water moderator and is traversed by horizontal tubes containing the fuel bundles and through which the coolant flows.

Cancer: A disease in which cells multiply uncontrollably in the body and interfere with the normal function of the organism.

CANDU: Canada Deuterium Uranium, describes a Canadian nuclear power reactor system using deuterium as moderator and natural uranium as fuel.

Chain reaction: A sequence of nuclear fission reactions in which the neutrons produced by the fissioning of one nucleus initiate fission in one or more other nuclei.

Cladding: The metal surrounding the uranium fuel in a fuel bundle. For CANDU fuel the cladding is a zirconium alloy.

CNSC: Canadian Nuclear Safety Commission, the federal body responsible for the regulation of nuclear activities in Canada.

Control rods: Rods consisting of a neutron absorbing material that are inserted into a reactor to change or stop the chain reaction.

Conversion: In uranium processing, the step in which uranium oxide (U_3O_8) is chemically converted to uranium hexafluoride (UF_6).

Coolant: A liquid or gas used to remove the heat from the core of a nuclear reactor. The most commonly used coolants are light water and heavy water.

Cosmogenic: Radioactive materials produced by cosmic radiation interacting with elements in the upper atmosphere.

Core: The central part of a nuclear reactor where the fuel is located and the chain reaction occurs.

Critical: In the nuclear context this describes a condition where a chain reactor is sustained.

Critical mass: The amount of a fissionable material required to just sustain a chain reaction. Its value depends on a variety of factors, for example, whether a moderator is present and the geometry of the arrangement of the material.

Curie (Ci): The (old) unit for expressing the rate of radioactive decay. 1 curie = 3.7×10^{10} disintegrations per second. One curie is the radioactivity of one gram of radium.

Daughter: Decay product. An older term used to describe the new nucleus that results when a given nucleus decays.

Decommissioning: The process of permanently removing a reactor from service which involves the safe dismantling of storage of its radioactive components.

Deuterium: The isotope of hydrogen, hydrogen-2, that has a nucleus containing one proton and one neutron. Also called heavy hydrogen, deuterium nuclei take the place of hydrogen nuclei in heavy water.

DNA: DeoxyriboNucleic Acid, long-chained double-helix molecules in the nuclear area of living cells that contain the genetic information of the organism.

Dose: General term for quantity of radiation received, also called exposure.

Electrolysis: The process of breaking up water molecules into hydrogen and oxygen by applying electricity.

Electron: A subatomic particle with negative charge that exists in the electron cloud surrounding the nucleus of an atom.

Element: A substance with atoms all of the same atomic number (same number of protons and electrons), but possibly different atomic weight (different isotopes).

Enriched uranium: Uranium in which the concentration of the isotope uranium-235 has been increased to greater than the natural value of 0.7% by weight.

Fast Reactor: A type of nuclear reactor without a moderator where the fission takes place at high neutron energies. Usually a fast reactor is a breeder reactor.

Fertile: An element that can be transformed by nuclear reactions into a fissile element.

Fissile: An element whose nucleus can be fissioned (split) by a neutron.

Fission: The splitting of a nucleus into smaller nuclei, usually two nearly equal ones. This is accompanied by the emission of neutrons and a significant amount of energy. Fission in a reactor is initiated by bombarding fissionable material, usually uranium-235 with neutrons.

Fossil fuel: A fuel such as oil, coal or natural gas originating from ancient vegetation which is burned to produce energy.

Fusion: A process in which two or more light nuclei combine to form a heavier nucleus with the release of energy. Fusion is the source of energy for the sun and the stars.

Gamma ray (γ): Energetic photon, electromagnetic radiation, emitted by the nuclei of some radioactive elements due to nuclear transitions.

Genetic effect: DNA mutation that can be transmitted to the next generation (offspring).

Gray: A unit of radiation dose, 1 joule/kg.

Half-life: The amount of time needed for half of the atoms in a quantity of a radionuclide to decay.

Heavy water: Water in which the natural concentration of deuterium, about 0.015% or one in 6600, has been increased to a high value, typically more than 95%. Heavy water is the moderator in CANDU reactors.

Hormesis: A theory indicating that small exposures to radiation have positive effects on biological systems.

IAEA: International Atomic Energy Agency. A United Nations agency with the purpose of promoting the peaceful uses of nuclear technology and preventing its diversion to weapons uses. It manages the world nuclear safeguards system.

Ion: Atom, molecule, or molecular fragment carrying a positive or negative electrical charge.

Ionizing radiation: Radiation that has enough energy to remove electrons from neutral atoms or molecules that it passes through, creating ions.

Irradiator: A device for exposing materials to radiation for a variety of scientific and industrial purposes.

Isotopes: Atoms of the same element, that is, having equal numbers of protons, but with different numbers of neutrons. Isotopes of the same element have the same chemical properties, because they have the same number of electrons.

Light water: Ordinary water as found in nature, compared to heavy water.

LLRW: Low-level radioactive waste

LWR: Light Water Reactor. The most common type of nuclear reactor, which uses light (ordinary) water as a moderator. There are two types of LWR: the boiling water reactor (BWR) and the pressurized water reactor (PWR).

MAPLE: Multipurpose Applied Physics Lattice Experiment. A new type of isotope production reactor. Two MAPLE reactors were built by AECL at Chalk River for MDS Nordion.

Mass number: The total number of protons and neutrons in the nucleus of an atom.

Megawatt (MW): One million watts, the usual unit used to express the electrical output of an electrical power plant, such as a nuclear reactor. One watt in the SI system of units is one joule per second.

MOX: Mixed Oxide fuel, reactor fuel consisting of a mixture of uranium and plutonium.

Moderator: The material used in a nuclear reactor to slow down, i.e. moderate, fast-moving neutrons so they will interact with the nuclei of fissile isotopes such as uranium-235 causing them to split, i.e. fission.

Molecule: The smallest unit into which a chemical compound containing one or more atoms can be divided and still maintain its chemical characteristics.

Natural uranium: Uranium with the isotopic abundance found naturally, i.e. 0.7 % U-235 with the remainder U-238.

Neutron: A subatomic particle with no electrical charge and a mass nearly equal to that of a proton. It occurs in the nucleus of all atoms except hydrogen.

Neutrino: A highly energetic particle with a very low mass produced in many nuclear reactions in the universe.

Noble gases: The chemically inert gases helium, neon, argon, krypton, xenon, and radon.

NPD: Nuclear Power Demonstration. This 22 MW reactor was the first in Canada to generate electricity (in 1962).

NRU: National Research Universal. A powerful research and isotope production reactor located at Chalk River, Ontario.

NRX: National Research Experimental. NRX was the first large research reactor at Chalk River, operating from 1947 to 1992.

Nuclear fuel cycle: All the steps necessary to use uranium to produce electricity including mining and milling of uranium, converting the uranium to a fuel, fissioning the uranium in a reactor to create power, and disposing of the waste. This term can apply more generally to other reactor fuelling schemes involving thorium, plutonium and other nuclei.

Nuclear radiation: Ionizing radiation (alpha, beta, gamma, neutrons) emanating from the nucleus of radioactive atoms as they undergo radioactive decay to a more stable form.

Nuclear reactor: A device in which a nuclear chain reaction is induced for power production, or other industrial and scientific purposes.

Nuclides: A general term used for species of atom nuclei; radionuclides are radioactive nuclides.

OPG: Ontario Power Generation, Canada's largest nuclear utility.

PET: Positron Emission Tomography, a medical diagnostic technique that employs the gamma rays that are created by the positrons emitted from the decay of certain radioactive nuclei.

Photon: A discrete amount of electromagnetic energy, or a quantum of light.

Plate tectonics: The study and description of the motions of the large plates which comprise the earth's surface.

Plasma: A gas consisting of electrically charged atoms and electrons.

Poison: Name (jargon) given to a substance that absorbs neutrons well, for example boron and gadolinium.

Positron: Similar to a beta particle or electron but with a positive instead of a negative electrical charge.

Ppm: parts per million

Primordial radionuclides: The radioactive nuclei formed at the birth of the universe.

Proton: A subatomic particle in the nucleus of an atom with approximately the same mass as a neutron and carrying a positive charge equal but opposite to that of the electron.

PWR: Pressurized Water Reactor, the most common type of electricity producing reactor. A PWR is one of the two general types of light water reactors. It has two coolant circuits. A primary coolant after passing through the reactor core transfers its heat to a secondary coolant that makes steam in a boiler-like device called a steam generator.

Radioactive decay: The spontaneous emission of radiation from the nucleus of a radioactive atom. It is the process by which a radioactive (unstable) nucleus becomes stable.

Radiopharmaceutical: A drug containing one or more radioactive elements.

Reprocessing: The chemical processing of spent reactor fuel to recover the fissile and fertile isotopes present in it.

Safeguards: A system of international treaties with monitoring designed to prevent nuclear materials from being diverted for weapons.

Sievert (Sv): A unit that measures the health risk of exposure of a living organism to ionizing radiation.

SLOWPOKE: A low-power research reactor designed by AECL that is safe for unattended operation.

SNO: Sudbury Neutrino Observatory, an experimental apparatus installed in a mine in Sudbury Ontario used to measure cosmic neutrinos, particularly those emitted by the sun.

Solar wind: The stream of particles and radiation emitted by the sun.

Somatic effect: Effect of radiation on the exposed individual and not on the offspring (genetic effect).

SPECT: Single Positron Emission Computerized Tomography, a more sophisticated version of PET.

Spent fuel: Nuclear fuel that has been used in a nuclear reactor to the point where it can no longer produce economic power. Also known as irradiated fuel or used fuel.

Tailings: The rock remaining after a mineral has been extracted.

Thermal reactor: A reactor in which the neutrons are moderated (slowed down) to permit fission at thermal (low) neutron energies.

Tokamak: A type of magnetic-confinement fusion reactor that uses current generated in a plasma to confine the plasma.

Transuranic: Elements having an atomic number greater than 92 (uranium). Transuranic elements are artificially produced and are radioactive.

TRIUMF: Tri University Meson Facility, a cyclotron at University of British Columbia.

Tritium: Hydrogen-3, a naturally-occurring radioactive isotope, also created in heavy water during reactor operation.

Wilson Cloud Chamber: A device for viewing particles by observing the ionization tracks they make in a vapour.

X-ray: Electromagnetic radiation used in medical and dental diagnoses because it penetrates matter. Although very similar to gamma rays, X-rays are created by energy transitions in the electron cloud that surrounds the nucleus.

ZEEP: Zero Energy Experimental Pile, the first nuclear reactor built outside the United States; it operated at Chalk River from 1945-1970.

Appendix A:

Radiation ABCs

THE ATOM AND ITS NUCLEUS

To understand nuclear radiation, we must understand the basic building block of all matter, the atom. An atom is the smallest unit of a chemical element that retains all the properties of that element. The word atom comes from the Greek word *atomos*, meaning indivisible, because at one time scientists thought that the atom was the smallest unit of matter.

Our solar system occupies a huge volume in space, most of which is empty. The atom is similar. Far inside a cloud of negatively-charged electrons is a tiny positively-charged nucleus. To gain a perspective, consider a baseball lying on the ground in the middle of the SkyDome in Toronto. If an atom were blown up in scale many billions of times, the dome would represent the outer fringe of electrons and the baseball would represent the atom's nucleus. It is from this tiny nucleus that nuclear power plants provide us with electricity. Now let us zoom in even closer and look at the nucleus.

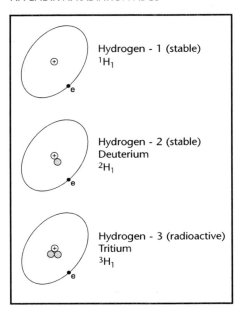

Figure A-1: Schematic of the three isotopes of hydrogen. Each isotope has one positively charged proton as indicated by the plus sign. Each also has one electron (designated by the e) orbiting the nucleus. Hydrogen-2, also called deuterium, has a neutron in its nucleus in addition to the proton. Hydrogen-3, also called tritium, has two neutrons. Tritium is radioactive and decays to helium-3 by emitting a beta particle.

Scientists have found that the nucleus is a bundle of two types of particles. The first type is called a proton and carries a positive charge, which is equal but opposite to the negative charge of an electron. An element is defined as an atom with a specific number of protons in its nucleus. There are 92 chemical elements that occur in nature and another dozen elements that have been created by humans. The number of protons is an unique characteristic of an atom, distinguishing which element it is. For example, an atom of oxygen always has eight protons, gold always 79, lead 82, uranium 92, and so on. Thus, the 92 natural elements are uniquely identified by having from 1 to 92 protons.

Neutrons, the second type of particle in the nucleus, are electrically neutral, as the name suggests. Neutrons and protons have almost the same mass, which is more than 1,800 times greater than the mass of an electron. The nucleus of a specific element always has the same number of protons, but it can have different numbers of neutrons. Thus each element has a "family" of atoms, called isotopes of that element, whose nuclei contain different numbers of neutrons. For example, hydrogen has three isotopes: each has one proton, which may be accompanied by zero, one, or two neutrons (see Figure A-1).

Under normal circumstances, the number of electrons orbiting around the nucleus is the same as the number of protons in the nucleus. Thus, since the charge on an electron is equal but opposite to the charge on a proton, the atom has no overall charge. The chemical behaviour of an element is determined by its electrons. This is the basic reason why the properties of, say, iron are so different from those of oxygen. You may remember the periodic table which lists the elements in rows and columns, where each column of elements has similar chemical properties. Chemical reactions between atoms are reactions involving their electrons.

Nuclide is the general name for a nucleus containing a particular number of neutrons and protons. A chart of the nuclides can be constructed by plotting the known nuclides on a grid with the number of protons along the horizontal axis and the number of neutrons along the vertical axis, in somewhat the same manner as the periodic table of chemistry. At the moment there are about 1,870 known nuclides.

To distinguish the different isotopes of an element, scientists have devised a notation using the atomic mass number, often just called the **mass number**, which is the sum of the number of neutrons and protons, and the **atomic number**, that is, the number of protons, which identifies which element it is. Uranium, for example, which has 92 protons, has a naturally occurring isotope with 143 neutrons. This can be denoted as:

$$\text{Mass Number} \longrightarrow {}^{235}U_{92} \longleftarrow \text{Atomic Number}$$

The superscript is the mass number, that is the number of protons plus neutrons (92 + 143 = 235), which identifies which isotope it is. The subscript is the atomic number, which identifies which element it is.

Note that the atomic number (92) is redundant since the "U" means uranium which is defined as the element with 92 protons in its nucleus. Another common way of writing this is uranium-235 or U-235, since the name uranium identifies the element and thus the number of protons.

Neutrons, being electrically neutral, play no role in chemical reactions. Thus, different isotopes of a given element always have the same chemical properties because they have the same number of electrons. However, they can have different nuclear properties. An element of particular interest is uranium, which in nature consists of a mixture of two isotopes; about 99.3% is uranium-238 (with 146 neutrons) and 0.7% is uranium-235 (with 143 neutrons).

QUANTUM MECHANICS

The world of the atom is governed by a physical theory known as quantum mechanics developed in the early 1900s. It has been highly successful in explaining and predicting many atomic and nuclear phenomena, and it is now firmly established as a primary theory in physics. Understanding quantum effects, however, can be difficult because they are often not what we would expect from our everyday experience. Concepts such as subatomic objects existing as both waves and particles, that the very act of observing these objects can change their properties leading to a degree of uncertainty about them, and that objects can be regarded as being distributed in space according to a system of probabilities, run counter to our normal experience. The good news is that you don't have to know about quantum mechanics other than to realize we are making great simplifications in this Appendix for purposes of clarity.

NUCLIDE STABILITY
AND RADIOACTIVE DECAY

Since like charges repel each other, why do the positively-charged protons in the nucleus not fly apart? The reason is that the nucleus is held together by a very powerful localized force called the nuclear binding force, which is stronger than the electrical force, but acts over a smaller distance. Neutrons help balance these opposing forces. If a nucleus has too many or too few neutrons to maintain this balance, it becomes unstable, or radioactive, and seeks to become more stable by a process called **radioactive decay**, also called radioactive disintegration.

Many nuclides are stable and will remain that way forever. Nuclear radiation, however, comes from those nuclei that are unstable, that is, radioactive, and most are. What does this mean? One way of looking at it is that a nucleus is stable for certain neutron numbers and unstable for others. Looking at a chart of the nuclides one can see that generally the number of neutrons in stable nuclides is somewhat greater than the number of protons, especially for the heavier elements. The pattern seen in the chart is that each element only has a limited number of isotopes (up to about 20 at most) with neutron numbers in a range around those of its stable isotopes. Of the approximately 1870 known nuclides only 290 are stable, the remainder (about 1580) have this property of instability or radioactivity.

Radioactive nuclides, or radionuclides, have excess nuclear potential energy and move toward stability by giving off energy, that is, by emitting radiation.

The primary types of nuclear radiation that are released during the decay process were named by the pioneering scientists after the first three letters of the Greek alphabet.

An alpha particle consists of two neutrons and two protons, essentially a helium nucleus. It is massive compared to other forms of radiation and, thus, can penetrate only short distances into matter. For example, it is stopped by a piece of paper or by skin. An alpha emission transforms the emitting element to a new element that has an atomic number that is two less; the mass number decreases by 4. An example of alpha decay, representing the alpha particle by the Greek letter α, is:

$$^{238}U_{92} \longrightarrow {}^{234}Th_{90} + \alpha$$

In this nuclear reaction a uranium-238 nucleus yields a thorium-234 nucleus and an alpha particle when it decays. The remarkable thing is that one element transforms to another, a process called transmutation. The primary objective of early alchemists was to find a way of producing gold from other elements. We now know transmutation is possible only through nuclear reactions.

A beta particle is the same as an electron, but instead of coming from the outer electron cloud, it is emitted from inside the nucleus. An electron is formed inside the nucleus when a neutron (n) converts to a proton (p) by emitting a nuclear electron (β).

$$n \longrightarrow p + \beta$$

In this process the number of neutrons in the nucleus decreases by one and the number of protons increases by one, again we have transmutation. An example of beta decay is:

$$^{59}Co_{27} \longrightarrow {}^{60}Ni_{28} + \beta$$

The atomic number increases by one but the total number of protons and neutrons stays the same, in other words, there is no change in the mass number. Depending on its energy, a beta particle can penetrate up to several centimetres into living tissue.

261

Positrons, denoted β^+, are like beta particles except they carry a positive charge. They are emitted from nuclei during the radioactive decay process when protons transform to neutrons.

$$p \longrightarrow n + \beta^+$$

An example of positron emission is:

$$^{22}Na_{11} \longrightarrow {}^{22}Ne_{10} + \beta^+$$

A positron is an antiparticle, an example of antimatter; it is seldom encountered because within microseconds it combines with an electron and both disappear, emitting two gamma rays. This phenomenon of positron annihilation is the basis of a diagnostic technique, positron emission tomography, widely used in nuclear medicine.

Gamma rays, denoted γ, are electromagnetic radiation emitted from nuclei, similar to visible light, radio waves, microwaves and X-rays. Unlike alpha and beta particles, gamma rays have no mass and no charge, only energy. Like all electromagnetic radiation, gamma rays travel at the speed of light. If only one or more gamma rays, and no other radiation, are emitted by the nucleus, then the change is called an isomeric transition, meaning the atomic and mass numbers remain the same. The original, higher-energy-level nucleus is said to be in an "excited state" and is indicated by an "m" beside the mass number. An example of an isomeric transition is in the element technetium (a widely used radioisotope in medicine):

$$^{99m}Tc \longrightarrow {}^{99}Tc + \gamma$$

Gamma rays are similar to X-rays but with higher energy. The primary difference is that gamma rays come from the nucleus, whereas X-rays arise from the atom's outer electron cloud. Like gamma rays, X-rays can penetrate relatively large distances into matter. Depending on their energy, gamma rays can penetrate up to a metre into concrete, and pass right through a human. It is important to stress that although X-rays cause the same health effects as nuclear radiation, they are an atomic effect, not a nuclear phenomenon.

Other transformations are possible inside the nucleus, for example, electron capture and internal conversion. Other particles may be emitted from the nucleus, such as neutrons, protons, and neutrinos. They will not be discussed further.

The stability of a nucleus depends largely on the mix of protons and neutrons. Except for hydrogen-1 and helium-3, the number of neutrons in a stable nucleus either equals or exceeds the number of protons. If an atom has too many neutrons it decays by emitting a beta particle, thereby converting a neutron to a proton. If the nuclide has too many protons, it decays by emitting a positron or (rarely) by "capturing" an electron in the nucleus, thereby converting a proton to a neutron.

Decay by emitting an alpha particle occurs only in the heavy elements (atomic number 83, bismuth, and higher). In all these cases, a new **daughter** nuclide, or decay product, is created. In the example of alpha decay used above, thorium-234 can be said to be the daughter of uranium-238. If the daughter is unstable, i.e. radioactive, it will also undergo radioactive decay. This process will continue until a stable configuration is reached.

RADIOACTIVE DECAY OVER TIME

Each radioactive nucleus decays independently, that is, the nucleus emits alpha, beta, and/or gamma(s) at random times. Although any one atom behaves in a random manner in terms of when it decays, the average decay behaviour of a collection of many radioactive atoms is actually predictable. Radioactive decay has been observed to always occur in an exponential manner.

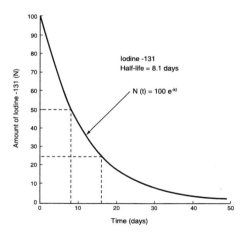

Figure A-2: Graph illustrating the concept of half-life for iodine-131. At the beginning there are 100 grams of iodine-131. After one half life (8.1 days) half of them have decayed and there are only 50 grams left. After 2 half lives (16.2 days) only a quarter (25) remain, and so on.

A useful parameter for characterizing the rate of decay of a particular nuclide is its half-life, which is defined as the time it takes for half the atoms to decay to daughter atoms, i.e. the time for the radioactivity to decrease by half. For example, the half-life of iodine-131 is 8.1 days. This means that if you started with 100 grams of iodine-131, after 8.1 days only 50 grams would be left (the other half would have transformed to another nuclide). After another 8.1 days, only 25 grams would be left, after another 8.1 days, only 12.5 grams would be left, and so on. If this is plotted on a graph it would take the form of an exponential curve, as shown in Figure A-2.

Half-life is an important concept. Every radioactive isotope or radionuclide decays following this exponential pattern until it becomes stable, that is, non-radioactive. Each radionuclide has its own distinctive half-life.

Some isotopes have very long-half lives; uranium-238, for example, has a half life of 4.5 billion years. It is convenient to use scientific notation to write very large numbers or very small numbers and a brief explanation of this method is included at the end of this Appendix.

Table A-1 lists the half-lives of some of the radionuclides that are discussed in this book. The first four nuclides listed in the table are naturally-occurring isotopes found in virtually all the rocks that make up the earth. Because they have very long half-lives, these nuclides have been around since the earth was first formed (4.5×10^9 years ago). We can conclude that the earth was much more radioactive in the past.

Table A-1
HALF-LIVES OF SOME RADIONUCLIDES

Uranium-238	4.5×10^9 years
Uranium-235	0.71×10^9 years
Thorium-232	14×10^9 years
Potassium-40	1.3×10^9 years
Hydrogen-3 (tritium)	12 years
Carbon-14	5730 years
Cobalt-60	5.3 years
Strontium-90	28 years
Cesium-137	30 years
Technetium-99m	6 hours

RADIOACTIVE DECAY CHAINS

Uranium and thorium, two heavy radioactive atoms that occur naturally in rocks and soils, have the remarkable characteristic that they do not decay directly to a stable element. Instead, their daughter products are also radioactive and in turn decay to their own daughters. This sequence continues through a chain of radioactive daughters until a stable one is reached. The uranium-238 decay chain is shown in Table A-2. Each isotope decays to the one below it in the table. For example uranium-238 decays to thorium-234 (by emitting an alpha particle), and the half life of this decay is 4.5 billion years. Then the thorium-234 decays to protactinium-234m ("m" means an excited state) by emitting a beta with half life 24 days, and so on. All the decays in the chain are happening simultaneously. It is seen that uranium-238 decays through 16 different daughters before reaching lead-206, which is stable.

Table A-2
THE URANIUM-238 DECAY CHAIN

ISOTOPE	HALF-LIFE
Uranium-238	4.5×10^{9} years
Thorium-234	24 days
Protactinium-234m	1.3 minutes
Uranium-234	2.5×10^{5} years
Thorium-230	7.7×10^{4} years
Radium-226	1622 years
Radon-222	3.8 days
Polonium-218	3 minutes
Lead-214	27 minutes
Astatine-218	2 seconds
Bismuth-214	20 minutes
Polonium-214	1.6×10^{-4} seconds
Thallium-210	1.3 minutes
Lead-210	22 years
Bismuth-210	5 days
Polonium-210	138 days
Thallium-206	4.2 minutes
Lead-206	stable

Similar natural decay chains exist for uranium-235 and thorium-232. In the thorium decay chain the gamma ray from thallium-208 is the radiation with the highest energy; in the uranium-238 chain it is the gamma ray from bismuth-214. A neptunium series once existed but is no longer observed, as the longest half-life in the series is only 2.2 million years, which is small compared to the age of our planet. In contrast, uranium-238 has gone through only one half-life since the earth was formed.

The complexity of these decay chains indicates how difficult it was for the early scientists to unravel them. It is a tribute to their patience and perseverance that they were able to accomplish this feat using what would now be considered primitive instruments.

UNITS OF RADIATION

To understand radiation in greater detail, we need to be able to talk about it quantitatively, that is, using numbers instead of generalities. To do this we must introduce the units in which radioactivity is measured.

The activity or amount of radioactivity in a substance is measured by the number of disintegrations, i.e., radioactive decays, that occur per second. One becquerel (Bq), is defined as one radioactive decay per second (this unit is named after Henri Becquerel who in 1896 was the first to detect radioactivity). The concentration of radioactivity in gases, liquids, and solids is usually expressed in becquerels per cubic metre, litre, and kilogram, respectively. Table A-3 gives an idea of the amount of radioactivity found in natural substances.

Table A-3
TYPICAL AMOUNTS OF
RADIOACTIVITY IN NATURAL SUBSTANCES

Radon-222 in air	30 Bq/m3
Radium-226 in drinking water	0.0003 Bq/L
Potassium-40 in soil	400 Bq/kg
Typical Human (total)	10,000 Bq

This table shows that there are typically about 30 radioactive decays per second caused by radon-222 in one cubic metre of air. Our bodies, for example, contain naturally-occurring radioisotopes with an activity greater than 10,000 Bq.

The concept of radiation dose, defined as the radiation exposure received by a living organism, and how it affects living cells is discussed in more detail in Chapter 4. However, the unit used to measure dose is briefly introduced here.

When ionizing radiation penetrates matter, it transfers energy to the substance. The unit gray (Gy) is used to measure the absorbed energy per unit mass. One gray corresponds to the deposition of one Joule of energy in one kilogram of matter. The gray is a large unit and more commonly the milligray (mGy) is used where the prefix "milli" means one thousandth, i.e., one Gy contains a thousand mGy.

The unit used to measure the effect of radiation on humans is the sievert (Sv), named after the pioneering Swedish clinical physicist, Rolf Sievert. The sievert is a measure of the energy per unit mass deposited by radiation in the human body (expressed in Gy) times the likelihood of that causing damage. Thus, the sievert takes into account the fact that different types of radiation cause different damage. For example, alpha particles cause more damage per unit of energy deposited than gamma rays. Doses received in real life are usually much smaller than a Sv, so the unit millisievert (mSv) is generally used.

Table A-4
TYPICAL RADIATION DOSES

Nuclear Power Plant Worker per year	0.1-1 mSv
Average Natural Radiation per year (Canada)	2.0 mSv
Typical Chest X-ray	0.3 mSv

The units described above are part of the internationally accepted system (Systéme International, SI) of units. The previous system of units still occurs in the literature, although they will not be used in this book. The previous units and their equivalence to the SI system are as follows:

> Activity: 1 **curie** (Ci) = 37 x 10^9 becquerel
> Absorbed energy: 1 rad = 0.01 gray
> Biological equivalent dose: 1 rem = 0.01 sievert

DETECTING RADIOACTIVITY

Radioactivity is very easy to measure because it can be detected at a distance, due primarily to the emission of gamma radiation. In comparison, the detection of non-radioactive substances requires sampling and chemical analyses at a laboratory, a much more costly and time-consuming process.

Some of the methods used to detect radiation are based on the fact that alpha and beta particles and gamma rays are all forms of ionizing radiation. That is, they knock atomic electrons out of their orbits when they pass through matter. The atoms that have been struck are no longer neutral; they each have an electrical charge and are called ions. The best known radiation detector is the Geiger counter, a hand-held instrument that uses a high voltage to detect the small electrical currents caused by ionization, even down to a single ionizing particle. Geiger counters are used, for example, by prospectors to search for uranium deposits and by

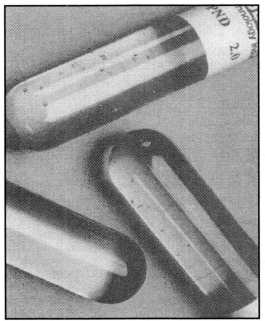

Bubble Technology Industries

Figure A-3: Measuring the radiation from neutrons is difficult but a Canadian company, Bubble Technology Industries, has come up with an ingenious solution. When neutrons strike a proprietary gel (lower tube without neutron exposure) they cause visible bubbles (upper two tubes). The bubbles are "frozen" in the gel and can easily be counted to determine the neutron field to which they were exposed. Bubble detectors have flown on many space missions and have been used in numerous other applications.

hospital safety workers to determine the extent of a radioactive spill and verify that the spill has been cleaned up.

A uniquely Canadian innovation, the "bubble detector" for measuring neutron exposure, is shown in Figure A-3. This ingenious but simple technology, discovered by AECL and commercially developed by Bubble Technology Industries of Chalk River, Ontario, has gained world-wide acceptance.

Because of their inherent "detectability", radioactive materials are easy to study, and the role of radioactive materials in the environment is generally far better understood than that of non-radioactive ones.

SCIENTIFIC NOTATION

In science and engineering one often has to deal with very large and very small numbers which do not easily lend themselves to the usual way of writing numbers. For example, the half life of uranium-238 is 4.5 billion years. This can also be written as 4,500,000,000 years. The scientific notation for this number is 4.5×10^9 which means exactly the same thing but in a shorthand way . Similarly, we can write one thousand (1,000) as 10^3 or 0.001 as 10^{-3}. This becomes more convenient as the numbers involved become very large or very small.

This notation also makes it easy to multiply numbers and provides a neat method for keeping track of the "zeroes". For example, one hundred (10^2) times ten thousand (10^4) is one million (10^6) this result is easily obtained by adding the exponents on the 10s, in this case 2+4 which equals 6. This is particularly useful when there are many very large and very small numbers to be multiplied. For example:

$$0.022 \times 11,000 = 2.2 \times 10^{-2} \times 1.1 \times 10^4 = 2.42 \times 10^2 = 242$$

Appendix B:

Nuclear Web Sites

American Nuclear Society (US)
- links to many other sites

www.ans.org

Atomic Energy of Canada Limited

www.aecl.ca

Bruce Power

www.brucepower.com

Cameco Corporation

www.cameco.com

Canadian Nuclear Association

www.cna.ca

Canadian Nuclear Safety Commission

www.cnsc-ccns.gc.ca

Canadian Nuclear Society

www.cns-snc.ca

Energy Probe

www.energyprobe.org

Greenpeace Canada

www.greenpeacecanada.org

Hydro Quebec

www.hydro.qc.ca

International Atomic Energy Agency

www.iaea.or.at

MDS Nordion

www.mds.nordion.ca

Natural Resources Canada

www.nuclear.nrcan.gc.ca

New Brunswick Power Commission

www.nbpower.com

Nuclear Energy Institute (USA)

www.nei.org

Ontario Power Generation

www.opg.com

Sierra Club

www.sierraclub.ca

Society of Nuclear Medicine

www.snm.org

Uranium Information Centre (Australia)

www.uic.com.au

Virtual Nuclear Tourist

www.nucleartourist.com

Wastelink
- more than 6500 links worldwide

www.radwaste.org

World Nuclear Association (UK)
- formerly Uranium Institute

www.world-nuclear.org

Acknowledgements

This project has taken several years to bring to fruition, and along the long road many people have given kind assistance. We would like to thank the Canadian Nuclear Society, MDS Nordion, and Zircatec Precision Industries Inc. for providing financial support to one of us (HT).

The manuscript has been improved enormously thanks to the contributions of many people who provided suggestions, illustrations and encouragement including:

Victoria Adams, Bob Andrews, B. Blann, Jim Blyth, Rachelle Bridge, Tami Brioux, Richard Bolton, Fred Boyd, Morgan Brown, Peter Brown, Jeffrey Carleton, Bill Clarke, Alan Cimprich, Ray Clement, Susan Copeland, Jerry Cuttler, Steve Dean, Phillipe Duport, William Evans, Sue Fletcher, Eva Gallagher, Chet Gray, Bruce Gullen, Tim Jackson, David Lisle, Bob Loree, Grant Malkoske, Donna McFarlane, David McInnes, Rankin MacGillivray, Mark McIntyre, David McLellan, Bernard Michel, Rita Mirwald, Lori Mondon, Eugene Martinello, Aniket Pant, Ed Price, Clair Ripley, Ben Rouben, Bruno Ruberto, Norman Rubin, Frank Saunders, Jen Skinner, Ken Smith, Patricia Smith, Terry Squire, Victor Snell, Mike Taylor, Paul Thompson, Patrick Tighe, Tom Timusk, Colin Webber, Gordon Yano.

We are also grateful to many, many others for their encouragement and support.

PHOTO CREDITS
The credit for each photo or figure is indicated on the side of the illustration. We would like to thank the following organizations and individuals for giving us permission to use their illustrations in this book:

Atomic Energy of Canada Limited, Ballard Power Systems, Bubble Technology Industries, Cameco Corporation, Canadian Nuclear Association, Canadian Nuclear Safety Commission, Canadian Centre for Magnetic Fusion, EFDA-JET, Electric Power Research Institute, Energy Probe, Hamilton Health Sciences, International Atomic Energy Agency, Joint European Torus, Lawrence Livermore National Laboratories, McMaster University, MDS Nordion, Mike Graston, Natural Resources Canada, New Brunswick Power, Nray Services, Ontario Power Generation, Pratt and Whitney Canada, Tourism Yukon, Tri University Meson Facility, US Department of Energy and Westinghouse.

Index